Light Manufacturing in Africa

Light Manufacturing in Africa

Targeted Policies to Enhance Private Investment and Create Jobs

Hinh T. Dinh
Vincent Palmade
Vandana Chandra
Frances Cossar

A copublication of the Agence Française de Développement and the World Bank

© 2012 International Bank for Reconstruction and Development / International Development
Association or The World Bank
1818 H Street NW
Washington DC 20433
Telephone: 202-473-1000
Internet: www.worldbank.org

1 2 3 4 15 14 13 12

ISBN: 978-0-8213-8961-4
eISBN: 978-0-8213-8974-4
DOI: 10.1596/978-0-8213-8961-4

Library of Congress Cataloging-in-Publication Data
Light manufacturing in Africa : targeted policies to enhance private investment and create jobs /
by Hinh T. Dinh ... [et al.].
  p. cm.
  Includes bibliographical references and index.
  ISBN 978-0-8213-8961-4 — ISBN 978-0-8213-8974-4 (electronic)
  1. Manufacturing industries—Africa, Sub-Saharan.   2. Manufacturing industries—Government
policy—Africa, Sub-Saharan.   3. Industrial policy—Africa, Sub-Saharan.   4. Africa, Sub-Saharan—
Economic conditions—Congresses.   I. Dinh, Hinh T., 1953–
  HD9737.A3572L54 2012
  338.4'7670967—dc23

                                                                                    2011045309

Cover photo by Arne Hoel.
Cover design: Debra Naylor of Naylor Design.

Africa Development Forum Series

The **Africa Development Forum series** was created in 2009 to focus on issues of significant relevance to Sub-Saharan Africa's social and economic development. Its aim is both to record the state of the art on a specific topic and to contribute to ongoing local, regional, and global policy debates. It is designed specifically to provide practitioners, scholars, and students with the most up-to-date research results while highlighting the promise, challenges, and opportunities that exist on the continent.

The series is sponsored by the Agence Française de Développement and the World Bank. The manuscripts chosen for publication represent the highest quality in each institution's research and activity output and have been selected for their relevance to the development agenda. Working together with a shared sense of mission and interdisciplinary purpose, the two institutions are committed to a common search for new insights and new ways of analyzing the development realities of the Sub-Saharan Africa region.

Advisory Committee Members

Agence Française de Développement
Pierre Jacquet, Directeur de 1ª Stratégie et Chef Économiste
Robert Peccoud, Directeur de la Recherche

World Bank
Shantayanan Devarajan, Chief Economist, Africa Region

Sub-Saharan Africa

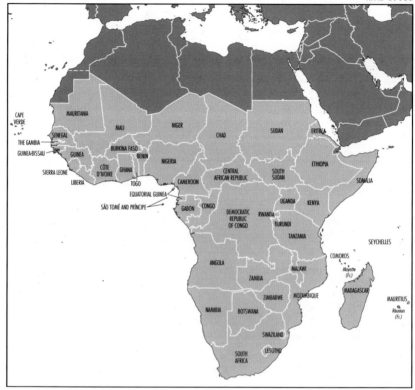

Source: World Bank.

Contents

Boxes

Figures

Tables

Foreword

The World Bank's strategy for Africa's Future recognizes the central importance of industrialization in Sub-Saharan Africa, and the consequent creation of productive jobs for Africans, which have long been a preoccupation of African leaders and policy makers. This book represents an attempt to address these issues.

The book stresses that, while the recent turnaround in Africa's economic growth is encouraging, this growth must be accompanied by structural transformation to be sustainable and to create productive employment for its people. For many African countries, this transformation involves lifting workers from low-productivity agriculture and informal sectors into higher productivity activities. Light manufacturing can offer a viable solution for Sub-Saharan Africa, given its potential competitiveness that is based on low wage costs and an abundance of natural resources that supply raw materials needed for industries.

With five different analytical tools and data sources, this book addresses the binding constraints in each of the five subsectors it covers: apparel, leather goods, metal products, agribusiness, and wood products. Ethiopia is used as an example, with Vietnam as a comparator and·China as a benchmark, and with insights from Tanzania and Zambia to draw out lessons more broadly for Sub-Saharan Africa. The text recommends a program of focused policies to exploit Africa's latent comparative advantage in certain light manufacturing industries—in particular, leather goods, garments, and agricultural processing. These industries could initiate rapid, substantial, and potentially self-propelling waves of rising output, employment, productivity, and exports that can push countries like Ethiopia onto a path of structural change of the sort recently achieved in both China and Vietnam. The timing for these initiatives is appropriate because China's advantage in these areas is diminishing due to steep cost increases associated with rising wages and non-wage labor costs, escalating land prices, and mounting regulatory costs.

This is the first research project based on the New Structural Economics, and it has five features that distinguish it from previous studies. First, the detailed work on light manufacturing at the subsector and product levels in five countries provide in-depth cost comparisons between Asia and Africa that can be used as a framework for future studies. Second, building on a growing body of work, the project uses a wide array of quantitative and qualitative techniques, including quantitative surveys and value chain analysis, to identify key constraints to enterprises and to evaluate firm performance differences across countries. Third, the findings that firm constraints vary by country, sector, and firm size led to a focused approach to identifying constraints and combining market-based measures and select government intervention to remove them. Fourth, the solution to light manufacturing problems cuts across many sectors and does not lie only in manufacturing. Solving the manufacturing inputs problem requires solving specific issues in agriculture, education, and infrastructure. African countries cannot afford to wait until all the problems across all sectors are resolved, thus making this focused approach valuable. Fifth, the study draws on experiences and solutions from other developing countries to inform its recommendations.

We hope that the report, part of the World Bank's knowledge creation and dissemination program, will contribute in a practical way to spur the creation of productive jobs in Sub-Saharan Africa.

Justin Yifu Lin
Sr. Vice President and Chief Economist
Development Economics
The World Bank

Obiageli Katryn Ezekwesili
Vice President
Africa Region
The World Bank

Acknowledgments

Under the guidance of Oby Ezekwesili (Vice President, AFR) and Justin Lin (Senior Vice President, DEC and World Bank Chief Economist) this report was prepared by a core team consisting of Hinh T. Dinh (Lead Economist, Team Leader and Lead Author), Vincent Palmade (Lead Economist and Co-Team Leader), Vandana Chandra (Sr. Economist), Frances Cossar (Junior Professional), Tugba Gurcanlar (Consultant), Ali Zafar (Sr. Economist), and Gabriela Calderon Motta (Program Assistant). The larger team responsible for the report includes, in addition to the above staff, George Clarke (Texas A&M International University), Kathleen Fitzgerald, Rita Lartey, Ying Li, Thomas Rawski (University of Pittsburgh), H. Colin Xu, Yutaka Yoshino, and Douglas Zeng (Washington); Marcel Fafchamps and Simon Quinn (Oxford University, England); Anders Isaksson (UNIDO, Austria), Mesfin Girma Bezawagaw, Nebel Kellow, Menbere Taye Tesfa (Ethiopia); Le Duy Binh, Pham Thai Hung, and Doan Hong Quang (Vietnam); Lihong Wang (China); George Gandye, Josaphat Paul Kweka, and Michael Ndanshau (Tanzania); Tetsushi Sonobe and Aya Suzuki (The National Graduate Institute for Policy Studies (GRIPS), Japan); the Global Development Solutions team (Washington); Yasuo Konishi, David Philipps, Glenn Surabian, Atdhe Veliu, John Weiss, Nebiye Gessese, and Christine Elbert; and Precise Consult team (Ethiopia). The work was carried out with the support and guidance of Zia Qureshi (Director, DECOS), Marilou Uy (Sr. Advisor, MDM and Former Director, AFTFP), Gaiv Tata (Director, AFTFP), Shanta Devarajan (Chief Economist, AFR), Greg Toulmin (Acting Country Director, Ethiopia), Ann E. Harrison (Former Director, DECVP), Asli Demirguc-Kunt (Director, DECVP), and Shahrokh Fardoust (Former Director, DECOS).

The report also benefited from key inputs from government officials in Ethiopia, Tanzania, Zambia, China, and Vietnam as well as from hundreds of private sector entrepreneurs interviewed in these five countries. In Ethiopia, we thank H. E. Ato Neway Gebre-Ab (Chief Economic Adviser to the Prime Minister), H. E. Ato Sufian Ahmed (Minister of Finance and Economic Development),

H. E. Ato Mekonnen Manyazewal (Minister of Industry), and H. E. Ato Tadesse Haile (State Minister for Industry) for their valuable comments. In China, we thank Messrs. Gao Fu and Li Qiang of the Ministry of Industry and Information Technology (MIIT), as well as officials from Jiangxi and Zhejiang Provinces, particularly Mr. Junming Wan, Ms. Huan Ren, and the Chinese Association of Development Zones for arranging the enterprise visits and for carrying out the quantitative survey in China. In Vietnam, we thank the Vietnam Chamber of Commerce, particularly Dr. Pham Thi Thu Hang, for organizing the enterprise visits and for executing the quantitative survey.

Throughout the preparation of this report, the team received valuable advice and guidance from an external advisory committee consisting of Yaw Ansu (African Center for Economic Transformation), Augusto Luis Alcorta (UNIDO), William Lewis (Founding Director, McKinsey Global Institute), Howard Pack (University of Pennsylvania), Jean-Philippe Platteau (Universite of Namure, Brussels), Kei Otsuka (The National Graduate Institute for Policy Studies (GRIPS), Japan), John Sutton (London School of Economics), Alan Gelb and Vijaya Ramachandran (Center for Global Development).

The peer reviewers are Ann E. Harrison, Ioannis N. Kessides, John Murray Mcintire, David McKenzie, Brian Pinto, Vijaya Ramachandran, and Tunc Tahsin Uyanik. In addition, the team has benefited from comments by Asya Akhlaque, Jean Francois Arvis, Paul Brenton, Hai-Anh Dang, Nora Dihel, Doerte Domeland, Michael O. Engman, Thomas Farole, Gary Fine, M. Louise Fox, Ian Gillson, Alvaro Gonzales, Michael Fuchs, Mombert Hoppe, Xiaofeng Hua, Leonardo Iacovone, Guiseppe Iarossi, Celestin Monga, Dominique Njinkeu, Paul Noumba, Gael Raballand, Ganesh Rasagam, José Guilherme Reis, Frank Sader, Marie Sheppard, Van-Can Thai, Papa Demba Thiam, Pham Van Thuyet, Volker Treichel, James M. Trevino, Dileep Wagle, and Chunlin Zhang.

The report was edited by a team headed by Bruce Ross-Larson at Communications Development. Earlier drafts were edited by Alison Strong and Paul Holtz. Financial support from MDTF, BNPP, PHRD, and UNIDO is gratefully acknowledged. The World Bank's Office of the Publisher (External Affairs) handled the production of the final printed volume under the guidance of Susan Graham, Stephen McGroarty, and Nora Ridolfi.

Abbreviations and Acronyms

AGOA	African Growth and Opportunity Act
COMESA	Common Market for Eastern and Southern Africa
DRC	domestic resource cost
EU	European Union
FDI	foreign direct investment
f.o.b.	free-on-board
GDP	gross domestic product
IGAD	Intergovernmental Authority on Development
KfW	German Development Bank
PPP	purchasing power parity
PTA	Preferential Trade Area for Eastern and Southern Africa
USAID	U.S. Agency for International Development

Light Manufacturing in Africa Study

MAIN REPORT
Volume I

Light Manufacturing in Africa (in print and on the Web) presents the findings of the Ethiopian study, with lessons for Tanzania, Zambia, and Sub-Saharan Africa.

BACKGROUND PAPERS
Volume II

Comparative Value Chain Analysis by Global Development Solutions, Inc. presents detailed value chain analyses for specific light manufacturing products for Ethiopia, Tanzania, Zambia, as well as China and Vietnam.

Volume III

Results from the Quantitative Firm Survey by Fafchamps and Quinn.

The Binding Constraint on Firms' Growth in Developing Countries by Dinh, Mavridis, and Nguyen.

Explaining Africa's (Dis) Advantage by Harrison, Lin and Xu.

Assessing How the Investment Climate Affects Firm Performance in Africa by Clarke.

Wages and Productivity in Manufacturing in Africa: Some Stylized Facts by Clarke.

Volume IV

Kaizen for Managerial Skills Improvement in Small and Medium Enterprises by Sonobe, Suzuki, and Otsuka.

All volumes can be found online at http://econ.worldbank.org/africamanufacturing.

Light Manufacturing in Africa: Targeted Policies to Enhance Private Investment and Create Jobs

After stagnating for most of the past 45 years, economic performance in Sub-Saharan Africa is at a turning point. Between 2001 and 2010 the region's gross domestic product grew at an average of 5.2 percent a year and per capita income grew at 2 percent a year, up from −0.4 percent in the previous 10 years. The reforms of the 1990s that focused on macroeconomic stability and liberalization began to gain traction. Between 2001 and 2010 net flows of foreign direct investment totaled about US$33 billion—almost five times the US$7 billion total between 1990 and 1999—and export growth was robust (World Bank 2011).

Experience elsewhere shows that this growth cannot be sustained without a structural transformation that lifts workers from low-productivity agriculture and the informal sector to higher-productivity activities. This transformation has yet to take place in Sub-Saharan Africa. The booming price of commodities (oil, cotton, metals, minerals, and others) that Sub-Saharan Africa mostly exports fueled a large part of the past decade's growth. Investment remains low in Africa—less than 15 percent of gross domestic product, compared with 25 percent in Asia—and more than 80 percent of workers are stranded in low-productivity jobs.

Labor-intensive light manufacturing led the economic transformation of many of the most successful developing countries, but it has not fared well in Sub-Saharan Africa. While China's emergence in the global manufacturing market since 1980 has resulted in a broad decline in the market share of all regions, the decline in Sub-Saharan Africa's share has been longer and deeper than most. Sub-Saharan Africa's share of global light manufacturing has continually declined—to less than 1 percent—and preferential access to U.S. and European Union (EU) markets has made little difference. Indeed, without structural transformation, Sub-Saharan Africa is unlikely to catch up with more prosperous countries like China and Vietnam, which were not very different from Sub-Saharan Africa in the 1980s.

In addition to increasing the productivity of medium and large formal firms, Sub-Saharan Africa has to raise the productivity and encourage the upgrading and expansion of small enterprises, mostly in the informal sector. Light manufacturing in Sub-Saharan Africa is characterized by a few medium-size formal firms providing products to niche or protected markets and by a vast number of small, low-productivity informal firms providing low-quality products to the domestic market. These enterprises provide low-paying jobs, little in foreign exchange earnings, and few productive employment opportunities for young Africans. Encouraging the productivity and expansion of small firms has not received adequate attention and will be addressed in this report.

This study draws on the following five analytical tools:

- New research based on the World Bank Enterprise Surveys
- Qualitative interviews with about 300 enterprises (both formal and informal) of all sizes in China, Ethiopia, Tanzania, Vietnam, and Zambia, conducted by the study team and based on a questionnaire designed by Professor John Sutton of the London School of Economics
- Quantitative interviews with about 1,400 enterprises (both formal and informal) of all sizes in the five countries listed above, conducted by the Centre for the Study of African Economies at Oxford University and based on a questionnaire designed by Professor Marcel Fafchamps and Dr. Simon Quinn of Oxford University
- Comparative value chain and feasibility analysis based on in-depth interviews of about 300 formal medium enterprises in the same five countries, conducted by the consulting firm Global Development Solutions, Inc.
- A study of the impact of Kaizen managerial training for owners of small and medium enterprises. This training, delivered to about 550 entrepreneurs in Ethiopia, Tanzania, and Vietnam, was led by Japanese researchers from the Foundation for Advanced Studies on International Development and the National Graduate Institute for Policy Studies.

A discussion of why we chose these countries and these analytical tools is presented in this report. Detailed results can be found online at http://econ.worldbank.org/africamanufacturing.

This study has five features that distinguish it from previous studies. First, the detailed studies on light manufacturing at the subsector and product levels in five countries provide in-depth cost comparisons between Asia and Africa. Second, building on a growing body of work, the report uses a wide array of quantitative and qualitative techniques, including quantitative surveys and value chain analysis, to identify key constraints to enterprises and to evaluate differences in firm performance across countries. Third, the findings that firm constraints vary by country, sector, and firm size led us to adopt a targeted approach to identifying constraints and combining market-based measures and selected government interventions to remove them. Fourth, the solution to light manufacturing problems cuts across many sectors and does not lie only in manufacturing alone. Solving the problem of manufacturing inputs requires solving specific issues in agriculture, education, and infrastructure. Fifth, the report draws on experiences and solutions from other developing countries to inform its recommendations. The report's goal is to find practical ways to increase employment and spur job creation in Sub-Saharan Africa.

Potential for Light Manufacturing: Creating Millions of Productive Jobs

Using new evidence, this report shows that feasible, low-cost, sharply focused policy initiatives aimed at enhancing private investment could launch Sub-Saharan Africa on a path to becoming competitive in light manufacturing. These initiatives would complement progress on broader investment reforms and could increase the share of industry in regional output and raise the market share of domestically produced goods in rapidly growing local markets for light manufactures. And as local producers increase the scale of their operations, improve the quality of their products, and accumulate experience with technology, management, and marketing, they can take advantage of emerging export opportunities. In Sub-Saharan Africa, as in China and Vietnam (which also experienced accelerated growth), policies that encourage foreign direct investment can speed industrial development and export expansion. The impact of isolated successes can be multiplied, as demonstrated by Ethiopia's recent foray into selling cut flowers in EU markets: a single pioneering firm opened the door to an industry that now employs 50,000 workers. The strategies proposed here could initiate a process with the potential to create millions of productive jobs.

Sub-Saharan Africa's potential competitiveness in light manufacturing is based on two advantages. The first is a labor cost advantage. In Ethiopia, for example, labor productivity in some well-managed firms can approach levels in China and Vietnam. At the same time, Ethiopia's wages are only a quarter of China's and a half of Vietnam's, and its overall labor costs are lower still. Sub-Saharan Africa's second advantage is an abundance of natural resources that supply raw materials such as skins for the footwear industry, hard and soft timber for the furniture industry, and land for the agribusiness industry. Institutional obstacles and unsuitable policies, however, have prevented local producers from taking advantage of certain resources. Timber costs are far higher in

Ethiopia than in China or Vietnam, leading Ethiopia to import Asian furniture despite Ethiopia's enormous unexploited potential to supply domestic timber, especially bamboo. This report envisions reforms that can unlock the potential of domestic resources like leather and bamboo to make a growing range of light manufactures competitive in the domestic market and, eventually, in global markets for labor-intensive products.

Is there room for Sub-Saharan Africa in the global market today? Yes, if Africa can exploit its opportunities soon. China dominates the global export market in light manufactures, and its competitive edge far exceeds that of low-income exporters that recently entered the global market. But the steeply rising costs of land, regulatory compliance, and especially labor (including both wages and benefits) in China's coastal export manufacturing centers have begun to erode these centers' cost advantage. This erosion will continue and probably accelerate in the coming years. New entrants have already begun to line up: Bangladesh, Cambodia, and China's interior provinces. The ongoing redistribution of cost advantages in labor-intensive manufacturing presents an opportunity for Sub-Saharan Africa to start producing many light manufactures, enhance private investment, and create millions of productive jobs.

Fortunately, the current global trading environment favors Sub-Saharan Africa if it can overcome key constraints in the most promising subsectors. Along with low labor costs and abundant resources, Sub-Saharan Africa enjoys duty-free and quota-free access to U.S. and EU markets for light manufactures under the Africa Growth and Opportunity Act and the Cotonou Agreement.

Are these advantages enough to offset Sub-Saharan Africa's generally low labor productivity relative to that of its Asian competitors? Yes, if appropriate supportive policies are implemented. This report draws inspiration from Asia, which demonstrates the enduring benefits of adopting policies early that facilitate competitive output and input markets; attract foreign direct investment to capitalize on the region's comparative advantage in low-wage labor; and provide technological, commercial, and managerial expertise. Like Asia, Sub-Saharan Africa could benefit from applying policies that accord free access to domestic and international markets for the inputs and outputs associated with light manufacturing and create conditions conducive to attracting foreign direct investment.

Case Study: Ethiopia

Using the five analytical tools described above and data from three African and two Asian countries, this study confirms that select African countries have the potential for light manufacturing. To realize this potential, these countries must overcome constraints that vary by country, subsector, and firm size.

Previous studies identified constraints in a long list of cross-cutting issues, including corruption, red tape, inadequate utilities, poor transport, poor skills, poor access to finance, and so on. In contrast, our detailed analysis points to a smaller, more specific, and sometimes new set of key constraints. Narrowing the analysis can make the reform agenda more manageable within the financial and human resource constraints of most African countries. The report presents an in-depth diagnosis of the constraints in five light manufacturing subsectors in Ethiopia: apparel, leather products, agribusiness, wood products, and metal products. We propose policy reforms to address these constraints based on the successes of other countries.

Ethiopia navigated the global economic crisis in 2008–09 better than many developing countries, encountering only modest declines in exports, remittances, and foreign investments, which have since recovered beyond their pre-crisis levels. Growth in exports and earnings, in conjunction with a relative slowdown in imports, has enabled foreign exchange reserves to rise. Overall inflation has dropped to single digits, mainly due to declining food prices, and growth has remained strong at about 8 to 9 percent a year since 2009. The Ethiopian government is committed to achieving continued growth within a stable macroeconomic framework in the context of the new five-year development plan (Growth and Transformation Plan 2010/11–2014/15, Ministry of Finance and Economic Development 2010). The plan's strategic pillars include sustaining rapid economic growth by promoting industrialization, enhancing social development, investing in agriculture and infrastructure, and strengthening governance and the role of youth and women.

Although each country's economy is unique—and several aspects of Ethiopia's governance, institutions, and political environment set it apart—there are enough common factors to make Ethiopia a good exemplar for a large group of Sub-Saharan African countries. Ethiopia has many natural resources that can provide valuable inputs for light manufacturing industries serving both domestic and export markets. Among its abundant resources are cattle, which can be processed into leather and its products; forests, which can be managed for the furniture industry; cotton, which can support the garments industry; and agricultural land and lakes, which can provide inputs for agroprocessing industries. Ethiopia has abundant low-cost labor, which gives it a comparative advantage in less-skilled, labor-intensive sectors such as light manufacturing. Ethiopia also shares several negative factors with other low-wage African countries, such as shortages of industrial land, poor trade logistics (particularly in landlocked countries), and limited access to finance.

This report does not claim that its Ethiopia-specific findings apply to all Sub-Saharan African countries, although, as discussed in part II, there are some commonalities among the three Sub-Saharan African countries studied in this report. Nevertheless the analytical approach applied to Ethiopia can be replicated in other African countries to derive specific diagnoses and propose solutions tailored to country circumstances. Detailed policy recommendations for Tanzania and Zambia will be available shortly. Replicating this study's methodology beyond our countries of focus will enable a rich analysis of the constraints to light manufacturing in other Sub-Saharan African countries and provide concrete policy recommendations to facilitate the growth of this sector across the region.

Apparel: Poor Trade Logistics

Given the ready availability of the technology and skills required to manufacture garments, there is obvious potential for domestic firms to increase their share in the domestic and global clothing markets. A significant and growing labor cost advantage, access to a state-of-the-art and well-located container port in Djibouti, and duty-free access to the U.S. and EU markets offer Ethiopia the opportunity to expand its apparel industry. Foreign direct investment can accelerate the process of ramping up production and exports. Ethiopia's potential for expanding its production of high-quality cotton enhances the potential benefits associated with expanded production of clothing. The binding constraint on Ethiopia's competitiveness in apparel has been poor trade logistics, which wipe out its labor cost advantage and cut it off from the higher-value, time-sensitive segments of the market.

Establishing a green channel for apparel at customs, providing free and immediate access to foreign exchange, reducing the cost of letters of credit, and setting up an industrial zone close to Djibouti would resolve the most important trade logistics issues. As in China and Vietnam, these reforms would position Ethiopia to attract outside investors to lead the industry. Competitiveness could be reinforced by developing a competitive textiles industry (Ethiopia produces high-quality cotton and has cheap hydro-energy). While Ethiopia's apparel sector currently generates only US$8 million in exports and 9,000 jobs, Vietnam has—with policies similar to those recommended in this report—achieved US$8 billion in exports and created 1 million productive jobs.

Leather Products: Input Shortages

Ethiopia has even greater potential in leather, which is more labor intensive than apparel. Italian shoe importers express the highest regard for Ethiopian leather. Modest, targeted reforms could enable Ethiopia's large animal herds to produce vast amounts of some of the best leather in the world. Furthermore, the penalty of poor trade logistics is less serious because leather products are less time sensitive than apparel. The immediate constraint is limited access to high-quality processed leather.

Enabling imports of high-quality processed leather would easily and immediately solve the acute shortage of leather in the industry. Removing restrictions on the exports of leather would increase incentives to invest in the Ethiopian leather supply chain. The challenge is to expand the commercial production and sale of high-quality animal hides. Facilitating access to rural land for good-practice animal farms (already written into Ethiopia's current development plan) will open the door to large-scale commercial herding enterprises. Promoting better breeds, controlling cattle diseases, and enabling the use of cattle as collateral would expand the capacity of small-scale operators to contribute to a larger supply of quality hides. Ectoparasites—a disease that affects Ethiopian hides—can be controlled by a modest program costing less than US$10 million a year. And improving trade logistics along the same lines as apparel would further enhance competitiveness. Vietnam—with a population similar to that of Ethiopia—created 600,000 productive jobs in the leather products sector by implementing policies similar to those recommended in this report.

Agribusiness: High Input Prices
As exemplified by the success of coffee and cut flowers, the potential for agribusiness lies in low wages, varied soil and climatic conditions, opportunities to increase yields on cultivated land, and large tracts of unused arable land. Ethiopia has the second largest dairy cattle population in Sub-Saharan Africa, behind Sudan and followed by Tanzania. As in other sectors, problems lie in the market for inputs (in agriculture), where distortions lead to low productivity and high prices. Fixed prices for several food items discourage farmers from increasing productivity. Issues in the seed and fertilizer markets also contribute to very low agricultural productivity—yield rates for wheat are generally less than 1 ton per hectare. Investor access to rural land is also difficult because of the confluence of traditional rights with state ownership of land. And neither agricultural land nor cattle can be used as collateral for loans.

To unleash the vast potential of agribusiness, the government should facilitate private sector investments in key input markets, such as fertilizers and hybrid seeds, and extend rural food safety nets to vulnerable urban dwellers as an alternative to price caps and export bans. In recent years, the government has encouraged these investments, and this trend should continue. The government is also advised to pilot in key locations the use of land and cattle as collateral and facilitate access to rural land for good-practice investors.

Wood and Metal Products: Land and Finance
Firms in the wood and metal products subsectors rely on expensive imports of wood and steel. Ethiopia has the natural resources to develop a competitive supply of wood, but its current supply is unreliable, unsustainable, and poorly organized. Even in the better-managed wood product firms, labor productivity is low—a worker produces 4.5 chairs a day in China and 1.9 in Vietnam, but

only 0.3 in Ethiopia. Ethiopia's steel subsector is also constrained. The price of steel is 30 percent higher in Ethiopia than in China due to poor trade logistics and high import tariffs.

The potential lies not in exports (at least initially) but in the growing domestic market and in the high weight-to-value ratio of finished wood and metal imports. The sector is dominated by smaller, mostly informal, firms with no large or exporting firms. For wood the government should facilitate access to rural land and financing for private wood plantations. For metals the cost of inputs could be reduced by cutting the 10 percent import tariff on steel and exploiting Ethiopia's proven reserves of iron ore. For both subsectors the government could support the most deserving enterprises by facilitating their access to skills, finance, and industrial land as part of "plug-and-play" industrial parks.

The Five Subsectors

In sum, although labor productivity is high among the few larger firms in four of the five subsectors, it remains low on average, especially for smaller firms that lack entrepreneurial, managerial, and technical skills. Access to industrial land and formal finance are also constraints in each of the subsectors. Smaller firms are constrained because the government gave larger exporters in certain sectors (including apparel and leather) preferred access to land and finance. Banks do not accept machinery or cattle as collateral, and restrictions on buying and selling make land difficult to use as collateral. Without easy access to land and finance, small firms remain small, low in productivity, and unable to upgrade technology or expand production.

Input policy issues are the most significant constraints (table 1). This finding is important because issues related to input industries tend to be overlooked in the reform agenda for light manufacturing. Indeed, Ethiopia (like many other African countries) should reform the many input industries where it has a comparative advantage—leather, wood, many agricultural products, and possibly apparel and steel—thanks to abundant natural resources, a favorable climate, and the potential for cheap hydro-energy.

Detailed policy recommendations, including the time frame for implementation, are presented in table 2. Many of these recommendations, consistent with the traditional agenda for investment climate reform, seek to promote competition and reduce transaction costs (such as improved trade logistics and lower import tariffs). In this case, however, the detailed subsector diagnostics and the cross-country comparisons reduce the number of policy recommendations and make them more specific. More importantly, the benefits of policy reforms are easy to identify. Some new critical issues have also emerged, such as the need to develop wood plantations. Even implementation of the "traditional" investment climate reforms can be dramatically improved by focusing,

Table 1 Constraints in Ethiopia, by Importance, Size of Firm, and Sector

		Input industries	Land	Finance	Entrepreneurial skills	Worker skills	Trade logistics
Apparel	Smaller	Important	Critical	Critical	Important	Important	
	Large	Important			Important		Critical
Leather products	Smaller	Critical	Critical	Critical	Important		
	Large	Critical			Important		Important
Wood products	Smaller	Critical	Important	Important	Important	Important	
	Large	Critical	Important	Important	Important	Important	
Metal products	Smaller	Critical	Important	Important	Important	Important	
	Large	Critical	Important	Important	Important	Important	
Agribusiness	Smaller	Critical	Critical	Critical	Important		
	Large	Critical	Critical	Important			

Source: Authors.
Note: Blank cells are not a priority.

at least initially, on the subsectors where they will yield the most benefits (such as trade logistics for apparel and leather products) and for which the country has a comparative advantage.

Implementing this policy agenda will require coordination across various government departments and agencies and a clear understanding of the objective of promoting the competitiveness of light industry manufacturers in the domestic and global markets. Before they commit resources, domestic and foreign investors need to see credible commitments from government pledging to complete the reform agenda. High-level support is a crucial ingredient in execution of the report's recommendations. We suggest that the Ethiopian government put in place a dedicated high-level team to develop and implement the proposed reform program.

The priorities and sequencing of policy reforms and interventions should adhere to three criteria. First, these measures should focus on sectors and subsectors demonstrating the most potential for comparative advantage and job growth. Second, measures should be the most cost-effective in the short and long runs, with the least fiscal impact. Third, implementation capacity and implications for governance and the political economy of policy reforms should be thoroughly assessed and used as a guide.

The report highlights the importance of industry-specific issues (such as the effects of cattle disease on leather, import tariffs on steel, and price caps on agricultural products) in addition to economywide issues (such as poor access to industrial land, lack of technical skills, and poor trade logistics). This raises the

Table 2 Proposed Policy Measures in Ethiopia

Sector	Short term (6–12 months)	Medium term (1–2 years)	Long term (2–5 years)
All sectors	(a) Remove import tariffs on all inputs for light manufacturing, including those that are destined for national and regional markets; (b) offer incentives for banks and other financial institutions to offer financing for machinery to well-managed firms	(a) Develop collateral markets to improve access to finance; for example, facilitate the use of machines, livestock, and land as collateral; (b) support programs such as Kaizen's managerial training on marketing and business strategy, production and quality management, and business record keeping; (c) improve the performance of trade between Addis Ababa and Djibouti, investigate and address issues relating to letters of credit, fuel costs, and competition among trucking companies; (d) harmonize and improve customs procedures by simplifying procedures and leveraging information technology	(a) Develop "plug-and-play" industrial zones to facilitate access to industrial land and basic infrastructure; consider setting up zones next to Djibouti; (b) develop the hard infrastructure (across boundaries) to support multimodal systems, which would entail rehabilitating the railway line between Addis Ababa and Djibouti (planned); (c) develop strategic partnerships along key trade corridors, including with Djibouti, to optimize port operations and minimize charges; (d) in partnership with the private sector, design technical skills development programs such as those at the Penang Skills Development Centre (Malaysia) that offer a variety of sector-specific short- and long-term certificate, diploma, and degree courses
Apparel	(a) Remove restrictions on exports of cotton; (b) guarantee the immediate availability of foreign exchange for apparel producers; (c) eliminate foreign exchange fee; (d) establish a green channel for apparel at customs; (e) negotiate lower handling fees with Djibouti	See the proposals for all sectors	See the proposals for all sectors
Leather products	(a) Remove restrictions on exports of leather and facilitate the import of high-quality processed leather; (b) guarantee the immediate availability of foreign exchange for leather producers; (c) eliminate foreign exchange fee; (d) establish a green channel for leather at customs; (e) negotiate lower handling fees with Djibouti	(a) Control cattle disease; (b) facilitate access to rural land for strategic investors in cattle through an inclusive and transparent process	Promote cross-breeding of cows

Wood products	Reduce taxes on legal wood	(a) Facilitate access to land and financing for sustainable private wood plantations of fast-growing species on degraded land close to the main urban centers; (b) facilitate access to rural land for strategic investors in wood plantations through an inclusive and transparent process	(a) Fight illegal logging; (b) enable carbon financing
Metallic products	See the proposals for all sectors	(a) Promote the exploitation of iron ore deposits; (b) conduct a feasibility study to assess the competitiveness of a domestic steel industry	See the proposals for all sectors
Agricultural products	Remove price controls on agricultural products	(a) Facilitate private sector investments in key input markets such as hybrid seeds and fertilizers (for example, by easing access to foreign exchange); (b) facilitate access to rural land for strategic investors in agricultural plantations through an inclusive and transparent process; (c) provide technical assistance to smallholders to connect with strategic investors	Promote cross-breeding of cows

Source: Authors.
Note: Proposals would benefit both small and large firms, with the exception of the promotion of iron ore deposits, which would only benefit large firms.

issue of how to prioritize and package policy reforms within and across sectors given that the government cannot tackle all issues at once. Specific reforms and economywide reforms need not be mutually exclusive. They can be complementary, and both are needed to move the economy forward. Priorities should be determined by cost-benefit analysis. Where hard choices are required, the government should give priority to the locations and industries with the greatest potential, taking into account the political cost of reforms. This approach would also help to expand the reforms through demonstration effects. China adopted this approach with land reforms (initiated in the Shenzhen Special Economic Zone), and Mauritius adopted it with labor reforms (first limited to exporters). Of course, such deliberately targeted reforms may be prone to mistakes and capture—hence the importance of a transparent and professional process to implement reforms and the political courage to correct mistakes.

African Competitiveness in Light Manufacturing and Possible Solutions from Asia

Although the specific nature and relative importance of the binding constraints and policy responses will vary by country, subsector, and firm size, the analysis in Tanzania and Zambia confirmed that most of the main constraints fall into the same six categories as in Ethiopia: input industries, trade logistics, access to finance, access to industrial land, worker skills, and entrepreneurial skills. Detailed policy recommendations on Tanzania and Zambia will be presented in an upcoming report.

To address the issues identified above, other successful developing countries have implemented several practical solutions, explored below.

Plug-and-Play Industrial Parks
In developing plug-and-play industrial parks, China gradually learned to address multiple constraints. Beginning in the late 1970s, special economic zones provided mostly foreign-owned firms with access to industrial land, port facilities, standardized factory shell buildings, worker housing, and duty-free import of materials and equipment for export production. Initial success encouraged the proliferation of industrial parks for domestic firms serving both home and overseas markets. Zone operations subsequently expanded to include training facilities and one-stop shops for business regulations. These initiatives considerably reduced financing costs and risks for the better-managed small firms, allowing them to grow into medium enterprises despite their inability to obtain bank loans. This is how China avoided the "missing middle" problem, where the size distribution of firms is characterized by either very small or very

large firms. The parks also contributed to the development of industrial clusters by making it easier for smaller firms to locate near larger firms, generating economies of scale and scope for Chinese industries. And housing workers next to plants reduced the labor costs for plant operators as well as the cost of living for workers.

Proximity first to Hong Kong SAR, China, and later to domestic ports that gradually developed into world-class cargo facilities, enabled China's special economic zones to help resolve trade logistics issues. Establishing the Shenzhen Special Economic Zone next to Hong Kong SAR, China, transformed a fishing village into a leading light manufacturing city of 8 million people in less than 30 years. The parks also served as the testing ground for difficult reforms: Shenzhen led China's adoption of a market-friendly land-lease system and many other new institutional and regulatory arrangements.

The key to success is intense competition and cooperation among firms inside and outside the parks. Most parks did not preselect specific light industries, instead letting market forces drive the formation of specialized clusters. The parks, including the factory shells and housing, are financed by local governments as well as the private sector. The local governments' share is often financed by bank loans (with the zone's real estate as collateral), repaid with the additional tax revenues from increased economic activity. To be successful, parks must be part of a package of interventions that address the most binding constraints limiting production and trade.

Key Input Industries

China and Vietnam reformed and supported key input industries so they could become competitive. Nationally, both China and Vietnam encouraged foreign direct investment for key inputs (as in machine manufacturers) and developed sustainably managed wood plantations and competitive agricultural sectors. They supported input and output markets through the provision of land and financing, as with the local government-managed Yiwu market in China, now the largest commodity market in the world. China and Vietnam also provided support and coordination along the value chains, while gradually reducing the trade restrictions protecting domestic input industries. In addition, China exempted export producers in special economic zones from paying taxes and duties on imported inputs; offered tax rebates to exporters outside the special economic zones; and gradually developed programs to provide information and technical assistance on inputs, technology, and suppliers to small and medium enterprises. The report shows that removing Ethiopia's import duty on inputs in the light manufacturing sector would cost no more than 2 percent of total tax revenue, which could easily be recouped by imposing an excise tax, if needed.

Trade Logistics

Both China and Vietnam relied on good trade logistics at the outset of their light manufacturing journey by creating manufacturing zones close to ports and then exempting firms in those zones from numerous domestic regulations and tariffs and import restrictions. This approach enabled China and Vietnam to import efficiently the inputs that domestic firms could not produce or supply competitively, allowing firms based in the new zones to gain access to export markets cheaply and quickly. Locating industrial zones next to world-class ports with efficient customs also made a difference, as shown in Shenzhen.

Emulating China and Vietnam will not be enough for Sub-Saharan Africa, especially in the many countries without direct access to a world-class port. African governments need to work together to pursue regional integration by improving trade logistics along key business corridors, especially governance and regulation at ports. To improve connectivity and increase competition, African countries should continue to develop the hard infrastructure to support multimodal systems combining trucking, railways, airways, and shipping. To harmonize and improve customs, countries should simplify procedures and leverage information and technology. And to increase competition among freight forwarders and shipping and trucking companies, countries should remove price controls and restrictions on foreign direct investment.

First Movers

Entrants in new industries typically face high costs and risks. This is especially true in Sub-Saharan Africa, where industrial structure and infrastructure are limited and regulatory and governance risks are high. But the prospect of single entrants serving as catalysts for the rise of competitive new industries is real, as illustrated by Ethiopia's rose industry, which began with a single firm and quickly grew to more than a dozen firms with direct employment exceeding 50,000 workers (plus further job creation among suppliers of transport, packaging materials, and so on). The rise of Ethiopia's rose industry began after the government facilitated land access for the first rose plantation, demonstrating the potential benefits of undertaking a limited, sharply focused policy intervention on behalf of promising start-ups.

In China initial efforts to promote the expansion of low-wage, labor-intensive production have given way to programs offering support to start-ups in more complex sectors, often involving skilled labor and high technology. Support may begin before firms move into the industrial zones, with site selection proposals, project reviews, and licenses for land use and construction. After moving in, firms may obtain technical assistance, technological upgrading, and access to market information through networking to guide the firm and the industry to become nationally competitive. Support to first movers can be provided on a one-time basis (to avoid rent seeking) and does not have to be expensive.

Governments can provide a matching-grant scheme to share the high risks of first movers, such as paying for part of a feasibility study. Government support does not need to target large foreign firms. Official efforts to disseminate basic knowledge about markets and suppliers alone may unleash new industries, as with Zambia's corrugated tin roof industry. Support to first movers should be open, transparent, and consistent, and all candidates should have access to the same benefits and information.

Entrepreneurial Skills

Our quantitative survey shows that entrepreneurs who receive technical assistance at start-up can deliver significantly better business outcomes. Ethiopia's Ramsay Shoe Factory shows the importance of technical assistance in raising the quantity and quality of output. Africa's multitude of small informal firms includes many operators with substantial entrepreneurial talent. The Kaizen study shows that lack of basic management skills limits the capacity of small firms to accumulate assets and expand (Sonobe, Suzuki, and Otsuka 2011). It also shows that, as in the case of Ramsay, access to modest and inexpensive training or technical support can open the door for some informal firms to prosper and expand into substantial generators of employment and, in some cases, exports. African governments should seek opportunities to invest in programs that can improve managerial and technical skills, especially among small-scale operators already working in high-potential subsectors. Governments should also help domestic firms to adopt and adapt existing technologies by providing targeted technical support and advice to owners and entrepreneurs. This support should encompass not only informal firms, but formal, large enterprises as well.

One way to help entrepreneurs is through Kaizen's managerial training in three modules: marketing and business strategy, production and quality management (including a brief introduction to workplace housekeeping techniques and other Kaizen activities), and business record keeping. Even a three-week training program can improve entrepreneurs' management practices and substantially increase their willingness to pay for the program. Foreign direct investment offers another way to expand the domestic stock of entrepreneurial skills. In China firms first benefited from the knowledge provided by overseas Chinese entrepreneurs and foreign managers at multinational branch plants. Firms then benefited from the transfer of know-how associated with foreign direct investment, which contributed to a new generation of domestic entrepreneurs.

Vocational Training to Improve Workers' Skills

Even with its low-skill workforce, Sub-Saharan Africa could become competitive in some light manufacturing sectors. In the apparel sector, for example, small numbers of managers and technicians can guide hundreds of workers. Specialists report that inexperienced workers can learn to operate sewing machines in no more than two weeks. For the longer term, upgrading to more complex production will require a better-trained workforce than is currently available. But the expansion of light industry need not await higher school enrollment and better-quality schooling. Industrialization can begin rapidly by targeting promising sectors with modest skill requirements—the objective of this report—and then adopting policy measures that contribute to lowering the training costs facing would-be private investors. Together with the private sector, governments can offer technical assistance to foster industry-specific vocational training for less-skilled workers. Such training could be offered to smaller firms in industrial clusters (including informal ones) and to larger firms in industrial parks.

Governments, in partnership with the private sector, can leverage publicly funded programs to turn out technicians who can operate or repair simple machines, read instructions, and use the Internet to communicate and search for information. In addition to good basic education, governments could design technical skills development programs such as those at the Penang Skills Development Centre (Malaysia), which offers a variety of sector-specific short- and long-term certificate, diploma, and degree courses for all levels of the workforce. Dedicated skills development schools have had huge payoffs in countries where they were directed at sectors in which the country had a latent comparative advantage.

Implementation Issues

The report emphasizes how important it is for African governments to focus their scarce resources on resolving the most significant constraints on light manufacturing industries with the biggest potential. Reform implementation should take into account five considerations:

- First, policy interventions should begin with pilot initiatives and be continually revised and updated. As the Chinese reform experience shows, where policy reforms are needed, pilot initiatives can be designed in selected rural, geographically isolated areas to ensure that lessons are learned before scaling up (Zhang, de Haan, and Fan 2010). Implementation should be decentralized as much as possible to increase proximity to the private sector, improve accountability, and foster competition between local governments.

- Second, not all efforts to support selected industries will succeed. The Asian experience shows that governments should be willing to drop failing policies. For this reason it is important to keep pilot initiatives small and put in place effective monitoring and evaluation mechanisms to review and improve ongoing interventions regularly.

- Third, the fundamental role of the government in private sector development is to foster competition. Ongoing patterns of development in light manufacturing in China and Vietnam show that competition is the most critical aspect of industrial development. In both countries, initial preference for state-run enterprises has given way to national policies that facilitate the growth of light manufacturing industries through open competition.

- Fourth, it is important for African governments to mobilize support from all partners—the private sector, nongovernmental organizations, and the donor community—in their structural transformation efforts.

- Fifth, one of the best ways for the government to facilitate robust private sector growth is to maintain a stable and conducive macroeconomic environment and ensure that natural resources are well managed. Most African countries have made substantial progress in this area in recent years, and these efforts should continue.

The fiscal impact and political economy of light manufacturing policy reforms should be assessed before implementation begins. It is often said that limited capacity and weak governance make it difficult for many Sub-Saharan African countries to design and implement the specific policies proposed here. But the proposed approach is not expected to create serious governance issues for four reasons:

- First, although some of the proposed reforms would reduce the rent of vested interest groups, the extent of policy intervention is smaller than the traditional big-bang approach, and any political impact is likely to be smaller. The benefits can be calculated up-front, and a deeper analysis of political economy issues can be conducted to reduce the losses of vested interest groups.

- Second, the proposed policy reforms would be consistent with the latent comparative advantage of the economy. If subsidies and other supportive policies target industries consistent with latent comparative advantage, the required subsidies will be small but will attract many new entrants into a competitive sector, making rent seeking less likely.

- Third, policy interventions should increase competition (for example, by reducing entry costs and risks) as well as the capacity of firms to compete. This is the opposite of what past, failed industrial policies did in Africa, such as policies protecting and subsidizing a few companies in industries in which the country had no comparative advantage.

- Fourth, while the broader agenda of good governance is important, African governments should not wait for improvements before supporting the private sector with focused policies to enhance investment and create more growth and more jobs. The East Asian experience shows that, with capable leadership that promotes development by encouraging private business and enabling market forces, a country can achieve impressive results even if the governance issues have not been addressed at all levels.

References

Ministry of Finance and Economic Development. 2010. *Growth and Transformation Plan 2010/11–2014/15 Volume 1: Main Text.* Addis Ababa: Government of Ethiopia. http://www.mofed.gov.et/English/Resources/Documents/GTP%20English2.pdf.

Sonobe, Tetsushi, Aya Suzuki, and Keijiro Otsuka. 2011. "Kaizen for Managerial Skills Improvement in Small and Medium Enterprises: An Impact Evaluation Study." Background paper (Light Manufacturing in Africa Study). Available online as Volume IV at http://econ.worldbank.org/africamanufacturing. World Bank, Washington, DC.

World Bank. 2011. *World Development Indicators 2011.* Washington, DC: World Bank.

Zhang, Xiaobo, Arjan de Haan, and Shenggen Fan. 2010. "Policy Reforms as a Process of Learning." In *Narratives of Chinese Economic Reforms: How Does China Cross the River?* ed. Xiaobo Zhang, Shenggen Fan, and Arjan de Haan. Hackensack, NJ: World Scientific Publishing.

Part 1

Setting the Stage

Chapter 1

Good Possibilities for Light Manufacturing in Sub-Saharan Africa

After stagnating for most of the past 45 years, economic performance in Sub-Saharan Africa is markedly better, suggesting a turning point (Arbache, Go, and Page 2008). Between 2001 and 2010 gross domestic product (GDP) grew 5.2 percent a year and per capita income grew 2 percent a year, up from –0.4 percent in the previous 10 years and exceeding growth in both Latin America and the Caribbean and in high-income countries (World Economic Forum 2011). The reforms of the 1990s, which focused on macroeconomic stability and liberalization, began to gain traction. After decades of relying mostly on donor finance and its own resources, Sub-Saharan Africa found a spot in the global market for foreign direct investment (FDI). Between 2000 and 2009, net FDI flows averaged about US$22 billion a year, more than five times the average of US$4 billion a year between 1990 and 1999 (UNCTAD 2011). And export growth was robust. These growth-related indicators suggest that Sub-Saharan Africa's economies, typically perceived as the last stronghold of traditional agriculture and underdevelopment in a globalized world, have turned a corner.

Structural Transformation

Sub-Saharan Africa's recent growth would mark the start of a new development trajectory if the attendant structural transformation, another indicator of sustainable growth, were equally visible. But it is not, as reflected in trade trends. The role of exports in the economy increased a little, to 32 percent during 2000–09 relative to 27 percent in the previous decade, but export diversification remained elusive. While export growth accelerated in the last decade, 73 percent of it was attributable to mineral exports, which spiked in a commodity price boom. Africa's share in the global economy also remained marginal. Between 1980 and 2008 its share of global exports stagnated at 1.3–1.6 percent, while East Asia's share increased from 3.3 to 14 percent (World Economic Forum 2011). Nor did Sub-Saharan Africa's investment growth, which should follow a pickup in FDI, materialize.[1] The macroeconomic reforms of the 1990s led to more sustainable fiscal policies, controlled inflation, and better managed debt. Some countries went further and addressed fundamental structural rigidities by divesting public sector activities, opening some government monopolies, such as telecommunications, to private participation, and reducing public sector borrowing from domestic banks to expand opportunities for the financing of private investment. But a surge in private investment did not ensue.

Without structural transformation, it is questionable whether Africa can create the millions of higher-productivity jobs needed to lift workers out of agriculture and the informal sector. Structural transformation was the foundation of labor productivity growth and prosperity in Asia. Millions of better-paying jobs were created with the emergence of labor-intensive sectors where productivity was higher than in agriculture and the informal sector.

A large, low-productivity agriculture sector that employs as much as 60 percent of the labor force is typical for a Sub-Saharan African economy (UNCTAD 2011). Including workers in the informal sector, close to 80 percent of all Sub-Saharan African workers are in low-productivity, low-income jobs. In light of the negative association between employment in agriculture and income per capita (figure 1.1), the only channel through which to create better-paying jobs is structural transformation to reallocate workers from low-productivity agriculture and the informal sector into more productive economic activity in

Figure 1.1 Employment in Agriculture and Income in Select Countries, 2008

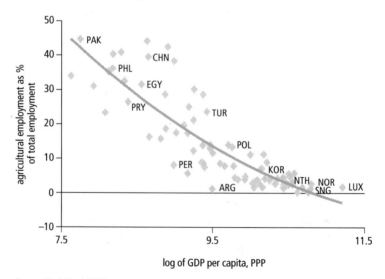

log of GDP per capita, PPP

Source: World Bank 2010.

manufacturing and services (McMillan and Rodrik 2011). Without this trans-formation, Africa cannot redress the prevailing market dualism that separates the thousands of small, mostly informal firms from the limited number of large, formal firms.

For the huge pool of workers in agriculture and the informal sector, labor-intensive manufacturing that requires skills not too different from their current ones could be a source of more productive employment. Agricultural incomes are invariably low, whether in coffee-producing Tanzania and Uganda—where about 75 percent of workers are in agriculture and 10–15 percent are in house-hold enterprises—or in cotton-producing Burkina Faso (nearly 85 and 10 per-cent, respectively; Fox 2011).

In addition to increasing the productivity of medium and large formal firms, Sub-Saharan Africa can derive great benefit from systematic efforts to raise the productivity of small enterprises, mostly in the informal sector. Light manu-facturing in Sub-Saharan Africa is characterized by a few formal medium-size firms providing products to niche or protected markets and by a vast number of small, informal, low-productivity firms providing low-quality products to the domestic market. These enterprises provide very low-paying jobs, little in foreign exchange earnings, and not much productive employment for aspiring young Africans.

The structure of most Sub-Saharan economies has not changed in the last half century. They continue to be dominated by agriculture (figure 1.2) or mining. While labor productivity in mining is undoubtedly high, it has limited potential for employment. The large service sector in most African countries comprises predominantly nontradable services, such as retail trade, where labor productivity is also low. In recent years, tourism seems to be growing well in some countries, but other tradable services have yet to emerge. The conspicu-ous absence of more productive, higher-wage employment options outside agriculture and the informal sector explains why, at the end of the first decade of the twenty-first century, incomes in Sub-Saharan Africa lag those of other developing countries by a considerable margin.[2] Worldwide, the average labor productivity is only US$17,530 in agriculture but US$38,503 in manufacturing (US$2,000 purchasing power parity; McMillan and Rodrik 2011).

Yet another metric of structural transformation is a shift in what Africa pro-duces and exports to the rest of the world. The emergence of new sectors and scaling up of more productive ones should elevate a country's position in the global market and should be measurable in a shift of its production or export basket from low-value primary products and services (such as household enter-prises engaged in petty trading) toward higher value added ones. This process has been slow to occur in Sub-Saharan Africa. Despite having the most open economies, as shown in the high share of trade in GDP, Sub-Saharan Africa saw its participation in global manufacturing production and exports decline

Figure 1.2 Economic Structure in Select Sub-Saharan African Countries, Various Years, 1960–2010

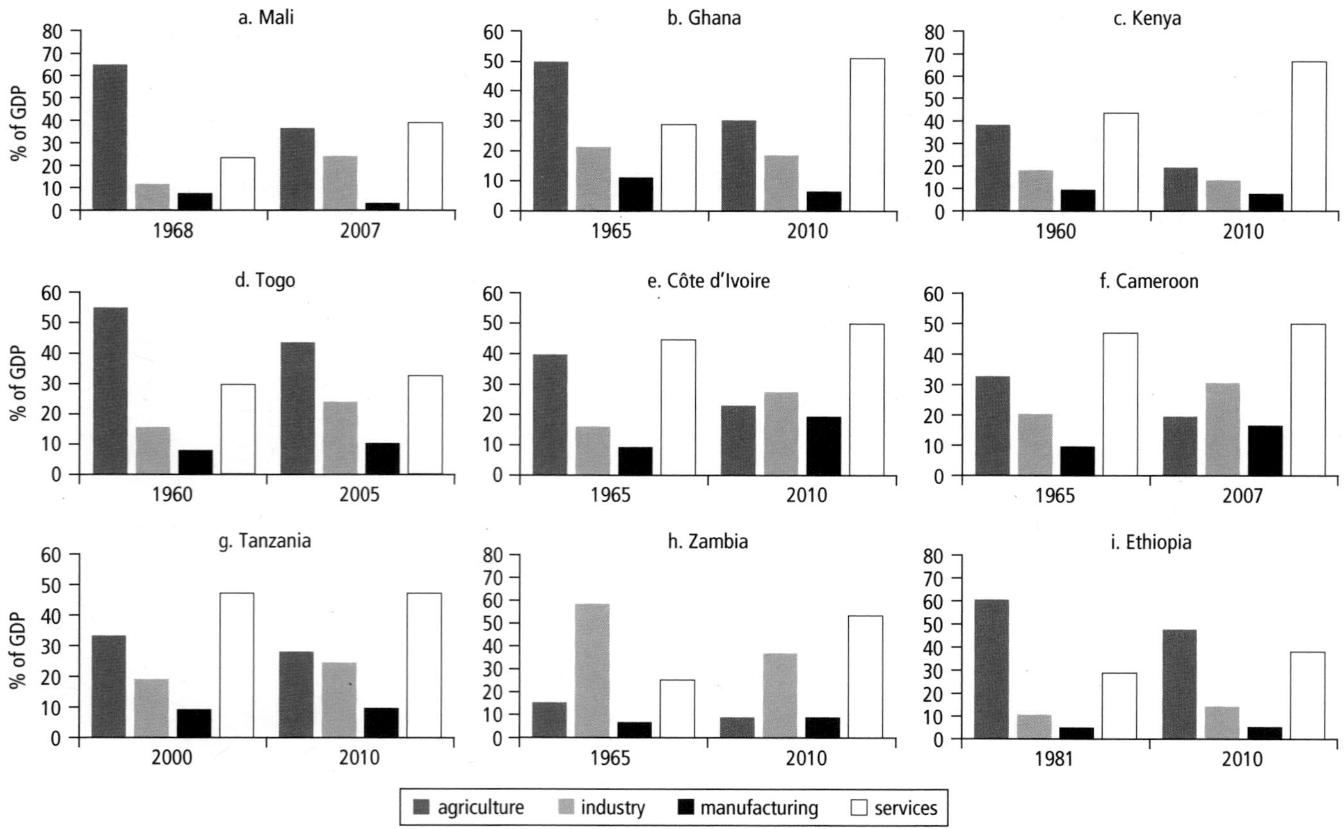

Source: Monga 2011; World Development Indicators data, various years.

Table 1.1 Monthly Wages in the Light Manufacturing Sectors of Five Countries, by Skill Level
US$

Product	China		Vietnam		Ethiopia		Tanzania		Zambia	
	Skilled	Unskilled	Skilled	Unskilled	Skilled	Unskilled	Skilled	Unskilled	Skilled	Unskilled
Polo shirts	311–370	237–296	119–181	78–130	37–185	26–48	107–213	93–173	—	—
Dairy milk	177–206	118–133	—	31–78	30–63	13–41	150–300	50–80	106–340	54–181
Wooden chairs	383–442	206–251	181–259	85–135	81–119	37–52	150–200	75–125	200–265	100–160
Crown cork	265–369	192–265	168–233	117–142	181[a]	89[a]	—	—	510[b]	342[b]
Leather loafers	296–562	237–488	119–140	78–93	41–96	16–33	160–200	80–140	—	—
Milled wheat	398–442	192–236	181–363	78–207	89–141	26–52	200–250	100–133	320–340	131–149
Average	305–399	197–278	154–235	78–131	77–131	35–53	153–233	80–130	284–364	157–208

Source: Global Development Solutions 2011.
Note: — = Not available.
a. Bottom of range; upper range not known.
b. Top of range; lower range not known.

These data refer to the cash wages paid to factory workers. Labor costs are not limited to cash wages; they also include employer contributions to pension plans, health and unemployment insurance, and other fringe benefits, as well as employer outlays on training, housing, recreation, and so on. Although systematic data are not readily available, nonwage labor costs in China are often high and seem likely to rise quite rapidly, perhaps more rapidly than cash wages. Since nonwage labor costs in many Sub-Saharan African countries are low and seem likely to remain so, expanding wage comparisons to include the full costs borne by employers seems certain to increase the potential cost advantages available to African firms entering labor-intensive light industries, both today and, especially, in the future.

In China, nonwage labor costs vary widely depending on location and type of firm. Urban manufacturers in the organized sector face substantial payouts. The labor costs of rural manufacturers and informal enterprises consist almost entirely of cash wages. Costs are expected to rise substantially in the future because the nonwage costs of registered urban enterprises are likely to rise and the range of firms expected to pay these costs is likely to expand.

There are several reasons for expecting the overall ratio of fringe benefits to cash wages to rise substantially in China in coming years. In the past, Chinese employers, especially in the south and in the private sector, have frequently ignored official regulations on wages, hours, and safety net contributions. Focusing on fringe benefit contributions, Banister (2005, 27) writes, "Noncompliance is rampant and penalties are rarely enforced." The same has been true in other dimensions of employment and compensation. Employers have used a variety of tactics to escape payment: underreporting earnings (Banister 2005, 28); simply ignoring laws and regulations; or substituting "contractors" (that is, workers supplied by local employment agencies often associated with the municipal labor bureau) for formal workers, a device that leaves the employment agency to cover (or, most probably, to ignore) contributions for fringe benefits.[5] Another big loophole arises from the presence of large numbers of migrant workers in Chinese urban factories; migrants are generally excluded from social programs aimed at registered urban residents.

Chinese authorities are gradually increasing their enforcement of existing regulations. In addition, recent legislation, particularly three laws enacted in 2008—the labor contract law, the employment protection law, and a statute governing mediation and arbitration of labor disputes—have strengthened the position of employees in labor-management relations, including migrant workers whose lack of urban residence permits had previously disqualified them from participating in municipal health, welfare, and pension programs. Academic summaries report that this legislative package will "increase labor costs" and "enhance labor costs of the enterprises," a view emphatically endorsed by employers during field visits to export-oriented manufacturers in Guangdong and Fujian (Zhao and Zhang 2010).

Access to benefits is expected to increase as a result of several trends:

- The 2008 labor legislation envisions a general formalization of labor relations, so that we should expect (a) tightening up of supervision and oversight, which will increase compliance among formal urban firms and also reduce the frequency with which local government agencies collude with employers in arrangements that effectively exclude substantial numbers of workers from participating in pension and safety net programs, and (b) increased pressure on the part of workers, including migrants, to benefit from arrangements mandated by laws and regulations. This will occur partly because of the new laws and partly because of market trends that increase workers' choice and thus bargaining power.
- There is a modest but growing tendency for some provinces and cities to lower (with the intent of eventually eliminating) long-standing barriers that restrict the migration of rural people into the cities and limit the eligibility of migrant workers to participate in pension, health care, education, and other systems designed to benefit registered urban residents.
- Chinese authorities have initiated a gradual expansion of pension, health, and safety net systems, initially established for the sole benefit of regular urban residents, to include growing participation by rural residents.

While the impact of these changes will be gradual rather than sudden and while some of the changes will not affect manufacturing costs, the cumulative impact on factory labor costs is likely to be substantial. For example, Yiping Huang (2010, 74, emphasis added) plausibly speculates, "If urban employers made social welfare contributions on behalf of their migrant workers, their *payrolls could rise by 35–40 percent.*" Chinese exporters, particularly in labor-intensive sectors like apparel, depend almost entirely on migrants to fill low- and semiskilled positions. A realistic appraisal of China's current political economy should include the expectation that the next 5–10 years will likely see a move to allow migrants working in urban factories to join at least some of the pension, health care, unemployment, and other programs described above. Any reform of this sort will further enlarge the substantial labor cost advantage available to Sub-Saharan African entrants into the light industry sectors that are the focus of this report.

A low-wage advantage alone does not guarantee Sub-Saharan Africa a comparative advantage in less-skilled labor-intensive manufacturing. As Sub-Saharan Africa competes with other low-wage regions (for example, South Asia), at least two other factors come into play. First, productivity is as important as wages in determining comparative advantage. Second, because wages and labor

Table 1.2 Labor Productivity in the Light Manufacturing Sectors of Five Countries

Labor productivity	China	Vietnam	Ethiopia	Tanzania	Zambia
Polo shirts (pieces per employee per day)	18–35	8–14	7–19	5–20	—
Leather loafers (pieces per employee per day)	3–7	1–6	1–7	4–6	—
Wooden chairs (pieces per employee per day)	3–6	1–3	0.2–0.4	0.3–0.7	0.2–0.6
Crown corks (pieces per employee per day × 1,000)	13–25	25–27	10	—	201
Wheat processing (tons per employee per day)	0.2–0.4	0.6–0.8	0.6–1.9	1–22	0.6–1.6
Dairy farming (liters per employee per day)	23–51	2–4	18–71	10–100	19–179

Source: Global Development Solutions 2011.
Note: — = Not available. Crown cork (bottle cap) production in Zambia is fully automated. Figures for wheat processing are from a sample of small enterprises in all five countries.

productivity vary across sectors, sector specificity is an important determinant of a country's comparative advantage in labor-intensive light manufacturing (table 1.2). Relative to East Asia, Sub-Saharan Africa's labor productivity is low except in *some* sectors in *some* countries. In sectors in which Ethiopia has a comparative advantage, also characterized by competition, good management can elevate labor productivity to a point comparable to that of the average firm in China or Vietnam.

Physical labor productivity in the three African countries falls within the range observed in Chinese and Vietnamese firms (although product quality may fall short of their standards, as occurs with polo shirts) except for wooden chairs and also Chinese flour milling, where the productivity figures shown in table 1.2 pertain to small operations located far below the domestic best-practice frontier.[6] When low wages are paired with the number of products a worker can produce in a day (labor productivity) in a well-managed firm, Ethiopia and Tanzania have a labor cost advantage. The production of one to seven pieces of footwear per employee per day in Ethiopia and four to six in Tanzania confirms this (table 1.2). On closer inspection, labor productivity is significantly higher in *well-managed* Ethiopian and even Zambian firms. And if other nonlabor-related constraints are lifted, it is high enough to restore the firms' labor cost advantage. For example, labor productivity in polo shirt production is 19 pieces per employee per day in a well-managed firm, underscoring Ethiopia's potential labor cost advantage in garments.[7] This also points to the importance of entrepreneurial skills, which determine whether a firm is well managed.

More comprehensive evaluations also validate Sub-Saharan Africa's comparative advantage in less-skilled labor. Findings of firm survey analyses for this study validate the value chain results. Across Africa, our study of wages and pro-

Figure 1.4 Labor Productivity per Worker in Select Countries, 2005

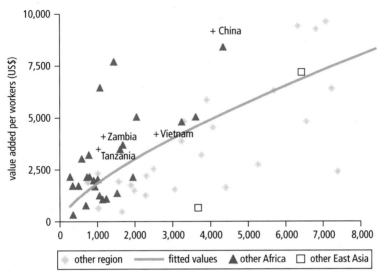

Source: Clarke 2011a.
Note: Calculations based on data from World Bank Enterprise Surveys. East Asia is China, Indonesia, the Philippines, Thailand, and Vietnam. Africa is Sub-Saharan Africa only. Data are for all Enterprise Surveys conducted since 2006 with at least 50 firms. Countries with GDP over US$8,000 are excluded. The fitted values line is from a log-log regression.

ductivity in manufacturing using the Enterprise Surveys shows that—considering per capita income (a proxy for institutional environment and the quality of physical infrastructure, which are shown by other studies to affect firm performance)—value added per worker (labor productivity) does not appear to be consistently lower in Sub-Saharan Africa than in other regions (Clarke 2011a). Indeed, more countries lie above the regression line than below (figure 1.4).[8] In a more detailed study, also from the World Bank's Enterprise Surveys (investment climate assessments) for formal firms, Harrison, Lin, and Xu (2011) examine the simultaneous effect of political, geographic, and business environments on firm performance. They find that, after controlling for the business environment, productivity, and sales growth, the formal manufacturing sector performs as well in Africa as in other countries at a similar level of income.[9] So, lower labor productivity in Sub-Saharan Africa's light manufacturing sector does *not* seem to be due to the characteristics of the workforce, such as weaker skills and unionization. And improving the business environment should go a long way toward restoring African competitiveness in light manufacturing.

In our quantitative survey of small and medium enterprises in Ethiopia, Tanzania, Zambia, China, and Vietnam, Sub-Saharan Africa's workers are found to be as productive as workers in East Asia.

Africa's Performance in Light Manufacturing

Weak competitiveness in nearly all light manufacturing industries and the marginal role of light manufacturing in exports and the overall economy sum up Sub-Saharan Africa's performance relative to that of other developing countries. Evidently, the celebrated growth rates of the last decade were not propelled by light manufacturing. Due to tougher global competition, producers in Sub-Saharan Africa cannot currently sell their products even in the domestic market and certainly not in the global market. The outcome? A slower or even arrested pace of structural transformation (as evident from Sub-Saharan Africa's exports to the world) and a widening gap in size and sophistication between Sub-Saharan Africa and other developing countries, particularly in Asia.

No Sign of Industrialization

A large part of Sub-Saharan Africa's higher GDP growth was propelled by price booms in agriculture and natural resources. Growth in manufacturing did recover from negligible values in the 1990s to about 4 percent a year during 2000–07, but had no effect on overall growth. Manufacturing contributes only about 8 percent of GDP in Sub-Saharan Africa today. Its gradual decline from nearly 11 percent in the 1980s confirms the dominance of agriculture and minerals and shows how far Africa is from becoming an industrial economy (table 1.3). Additional evidence of limited industrial progress comes from trade statistics and domestic production data confirming that many garments—as well as other light manufactures that Africa produced in the 1990s for domestic consumption—are now imported.

By contrast, East Asia and South Asia, the two fastest-growing developing regions, have enjoyed even faster growth of manufacturing, widening the gap with Africa. Between 1990 and 2007 East Asia sustained growth of about 10

Table 1.3 Level of Industrialization, by Region, 1960s–2000s
manufacturing as a % of GDP

Decade	East Asia and the Pacific	Europe and Central Asia	Latin America and the Caribbean	Middle East and North Africa	South Asia	Sub-Saharan Africa
1960s	24.8	—	25.6	—	14.2	9.4
1970s	31.5	—	26.5	—	15.7	10.1
1980s	31.5	—	26.5	12.3	16.1	10.7
1990s	30.4	21.3	19.6	14.2	16.1	10.8
2000s	31.1	18.5	18.0	12.1	15.7	8.5

Source: World Bank 2010.
Note: — = Not available.

percent a year, expanding the share of manufacturing in the economy to about 30 percent and laying the foundations for a strong industrial base. South Asia's manufacturing has grown 7 percent a year since 2000.

Until liberalization in the 1980s, Sub-Saharan Africa's markets were shielded from global competition, and many products were produced domestically. But in the last two decades, imports of even simple consumer products took over in African markets. It is now well established that competition breeds innovation, weeds out inefficient firms, and encourages efficient ones to upgrade, innovate, and compete with new entrants, thus leading to an overall more competitive sector. But Sub-Saharan Africa's manufacturing sector has defied this trend.

Why? Mainly because at the start of liberalization, the playing field was not level, and no one—not policy makers, academics, or donors—paid enough attention to light manufacturing. African nations made little effort to smooth the path for would-be private entrepreneurs, neglecting lessons from the post–World War II experience of numerous formerly low-income economies—including Japan; Taiwan, China; Hong Kong SAR, China; Singapore; the Republic of Korea; Thailand; Malaysia; China; and Vietnam—that have leveraged the achievements of private sector manufacturers to promote national prosperity. So Sub-Saharan African industries marked time, while Asian industries raced ahead, especially in apparel, footwear, and furniture. After trade liberalization, most African producers could not compete with cheaper and better-quality imports, especially from China, which currently dominates the global market for light manufactures.

Worse yet, liberalization of input markets in Sub-Saharan Africa, when it occurred, typically lagged behind product market liberalization. In some cases, crippling restrictions on input markets remain in place even today, as is evident in Ethiopia's leather industry. The lack of competition sustains high input costs, which further erode the competitiveness of the final product. Today, most African light industries are not competitive globally, and African producers struggle to compete with foreign producers to sell even milk, poultry, apparel, and shoes in the domestic market. But the best-managed firms in some sectors offer a glimmer of hope.

Manufactured Exports

East and South Asia are good benchmarks for evaluating the contribution of light manufacturing to the transformation of Sub-Saharan Africa's export basket from primary commodities to other products. Manufactured exports were pivotal in the catch-up of low-income East Asia with the middle-income East Asian Tigers in the 1980s and in the latter's catch-up with Japan in the 1970s (Lin 2011). Invariably, light manufacturing preceded heavy industry. The East Asian model of export-led catch-up is notable for the size of the boom in manu-

Table 1.4 Share of Manufactures in Total Exports in Select Countries in Africa and Asia, 1990–94 and 2005–09
% of total exports

Country	Manufactured exports		Light manufactures	
	1990–94	2005–09	1990–94	2005–09
Bangladesh	84	93	81	91
China	81	90	56	35
Ethiopia	22	13	10	9
Indonesia	27	38	19	16
India	51	54	41	29
Cambodia	22	90	21	89
Lao PDR	33	34	32	22
Pakistan	81	81	73	71
Tanzania	13	12	10	6
Vietnam	23	58	21	43
Zambia	3	6	2	3

Sources: World Bank 2010; COMTRADE SITC, 2–3 digit classification.

factured exports and the dramatic change in the composition of exports, showcased here by China and Vietnam.

Between 1990–94 and 2005–09 the share of manufactured exports in Chinese exports grew from 81 to 90 percent,[10] and the share in Vietnamese exports grew from 23 to 58 percent (table 1.4). In recent years the share of light manufactures in Chinese exports declined, as China diversified into more sophisticated, high-tech products. But in Vietnam the share expanded, signaling the sector's enormous potential in low-income countries. East Asia, however, is not the only region that leveraged export-led growth to become more prosperous, as Bangladesh, Nepal, Pakistan, and Sri Lanka show (table 1.4). In 2005–09 the corresponding shares were just 9 percent in Ethiopia, 6 percent in Tanzania, and 3 percent in Zambia.

Growth in light manufacturing does not have to be tied only to exports, at least not in the short term. Sub-Saharan Africa's domestic markets may be able to absorb increased production of some simple, low-price, import-competing products such as garments, furniture, and processed food. But in the medium term the larger scale of production and size of the market will demand an export-led model. Can Sub-Saharan Africa achieve export-led growth?

The case for export-led growth in developing countries has been well established (see World Bank 2009; Harrison and Rodríguez-Clare 2010; Chenery 1980). And recent evidence indicates that there may indeed be a case for growth led by exports of light manufactures in Sub-Saharan Africa (Harrison, Lin, and

Xu 2011). More outward-oriented countries have grown faster, although establishing the direction of causality is difficult (Harrison and Rodríguez-Clare 2010). Developing countries also need to export to get the resources to import skills and technology to move up the value chains. And in resource-based industries there are learning effects from being exposed to global competition.

The astounding contrast in the transformations of their export baskets two decades apart is clear testimony to the widening of the competitiveness gap between East Asia and Sub-Saharan Africa. In less than three decades, China transformed its export basket from gas, petroleum, live animals, and a few low-tech manufactures such as fabrics and footwear to the most technologically sophisticated medium-tech and high-tech products. The export share of the top 10 products in China's export basket did not shrink with the emergence of more light manufactures. Instead, it rose from 20 to 27 percent, reflecting the primacy of light and heavy manufactures. Footwear was the only common product that China exported in the 1980s and 2000s (table 1.5).

Table 1.5 exaggerates the sophistication of China's current export basket because the contribution of Chinese factories to digital data-processing

Table 1.5 Top 10 Non-Oil Exports of China, 1980–84 and 2004–08

1980–84			2004–08		
Product	Level of technology	%	Product	Level of technology	%
Cotton fabrics: woven and unbleached	Low tech	3.1	Complete digital data-processing machines	High tech	5.0
Gas and oils	Resource-based	2.8	Peripheral units: control and adapters	High tech	3.5
Linens, furnishing, and textiles	Low tech	2.8	Parts and accessories	High tech	3.2
Cotton fabrics: woven and dyed	Low tech	1.9	Television, radio broad-casting, and transmitters	High tech	3.1
Basketwork, brooms, and paint rollers	Low tech	1.9	Parts and accessories for machines	High tech	3.1
Footwear	Low tech	1.5	Footwear	Low tech	2.2
Fabrics: woven and synthetic fibers	Medium tech	1.6	Sound recorders, video recorders	Medium tech	2.1
Women and infants outerwear and textiles	Low tech	1.5	Electronic microcircuits	High tech	2.0
Fine animal hair, not carded or combed	Primary	1.5	Children's toys, indoor games, and so on	Low tech	1.9
Swine, live	Primary	1.5	Outerwear knitted or crocheted	Low tech	1.9
Total share		20.0	Total share		27.7

Source: COMTRADE, SITC 2–3 digit.

machines and other "high-tech" products often consists mainly of assembly and packaging. Even so, the rising share of "high-tech" products in China's export basket demonstrates the capacity of Chinese firms to capture substantial (and growing) shares of the value chains for products like desktop and tablet computers, smart phones, and printers that enjoy steep growth in demand and therefore contribute to China's rapid expansion of manufacturing output, employment, and exports.

In Sub-Saharan Africa, unlike Asia, the same products have been exported for more than three decades. Of the five Sub-Saharan countries listed with their top five exports, only Nigeria and Benin have expanded into low-tech exports (figure 1.5). Expanding the list shows that six of the top nine exports retained their place in Sub-Saharan Africa's export basket between 1980 and 2008. None of these were manufacturing items in 1980. By 2008 three products—iron ore, fuels, and tea—were replaced with diamonds, wood, and tobacco. Despite ample wood resources, as late as 2008, most of Sub-Saharan Africa was not processing simple sawlogs into furniture or roasting coffee for export.

Figure 1.5 Top Five Exports from Select Economies in Sub-Saharan Africa and Asia, 1980 and 2009

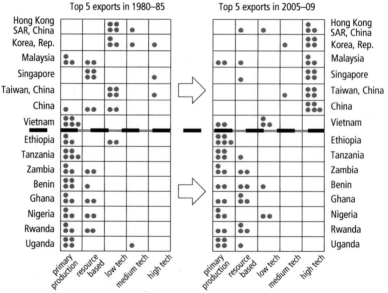

Source: PRMED Export Diversification webtool (based on COMTRADE statistics and adaptation of UNIDO's technology definition).
Note: Data for Ethiopia is 1990–95.

The Dual Industrial Structure in Africa

Overwhelming evidence from new research, including this study, indicates that the constraints on firms vary by size, so a one-size-fits-all approach is likely to be ineffective. Small and large firms need to be treated separately, with the eventual goal of integrating them into comprehensive value chains. In light manufacturing, in particular, a prerequisite for exporting today is having the capability to fulfill large orders competitively (price and quality) and quickly. Both require tapping into scale economies associated with labor-intensive, assembly-line production chains—that is, large firm operations. By definition, smaller firms cannot do this.

There are very few medium or large companies, and those that survive struggle to compete with imports. The striking paucity of medium and large firms explains immediately why Sub-Saharan Africa's light manufacturing cannot grow or chart an export-led growth trajectory. With the population of light industry enterprises skewed toward the proliferation of small firms, light manufacturing is constrained from playing a larger role in the domestic economy and virtually barred from exports. The small number of medium and large firms inhibits competition, discourages large new entrants, including would-be foreign investors, and stunts the sector.

It is not clear why there is such a glaring absence of large firms. Our study finds that firm growth[11] in a sample of small and medium producers of garments, leather products, processed foods, wood products, or metal products has been as robust within Ethiopia, Tanzania, and Zambia during 2008–10 as in Asia (Fafchamps and Quinn 2011). There is a substantial overlap in growth across the firms in Sub-Saharan Africa and East Asia—most Chinese firms in this sample did not grow much faster during 2008–10 than firms in Africa or Vietnam.

One explanation for the near absence of large firms in Sub-Saharan Africa's light manufacturing pertains to the skills required to organize and manage medium and large firms. The capabilities of small entrepreneurs are not adequate for graduating from the typical small enterprise into the very different population of mid-size manufacturers, which need in-depth industry knowledge and experience in managing a certain scale of operation (Sutton and Kellow 2010). Söderbohm (2011) confirms this finding: "The striking fact is that small manufacturing enterprises almost always stay small ... there is a huge (10-fold) gap in value added per head between manufacturing firms employing more than 50 workers and those employing fewer than 10 workers."

Another explanation could be that Asian governments facilitate the process that small entrepreneurs need if they are to grow into medium and larger firms. Firm size in Sub-Saharan Africa is positively associated with the stock of capital, machinery, and land, although information on land is sparse (Fafchamps and

Quinn 2011). That association may explain why firms in the industrial parks in China can overcome these hurdles and gain access to these factors, exploit scale economies through modern technology that facilitates assembly line production, and grow into larger firms. The qualitative interviews of firm owners validate this hypothesis.

A closer look reveals that Ethiopia, Tanzania, and Zambia produce a variety of simple manufactured goods from metals fabrication to agribusiness. But these goods are produced by a very large, dynamic informal sector that caters to the low end of the domestic market, generally with lower-quality goods that have no direct competition from imports. The informal firms start out small and take advantage of lower wages and lower costs of doing business (avoiding taxes and regulations), but they remain small due to their lack of other factors of production. Their labor productivity cannot match that of larger Chinese firms.

Without exception, the owners of informal firms do not have access to the land needed to expand the scale of production. So, while they do not need finance to start-up, they cannot grow to the size of an East Asian firm. The lack of land ownership precludes the use of land as collateral to obtain financing to purchase better machines and increase productivity. Clearly, attention to firm size will be crucial to any policy interventions targeted at jump-starting light manufacturing in Sub-Saharan Africa.

A very large number of small, mostly informal, firms engage in low-productivity work. The vast majority of firms in Sub-Saharan Africa are small, with many owned and operated by household members, mostly in the informal sector.[12] In Zambia, for example, about 84 percent of workers are in informal enterprises (World Bank 2011). In many countries in the region, wages are far lower in the informal sector than in the formal sector. Workers in large privately owned firms in Ghana and Tanzania earn more than twice as much as similar workers in small firms and self-employed persons (Rankin, Sandefur, and Teal 2010). The pattern is similar in Zambia, where sales and labor costs per worker are both very low among small and medium enterprises. The average monthly labor costs per worker were about US$120 a month in 2009 for the large, formal firms in urban areas (qualitative interviews), US$95 a month for registered small and medium enterprises in urban areas, and only US$43 a month for unregistered small and medium enterprises in urban areas (Clarke 2011a).

The implications of this split are clear. Low wages are a clear signal of the low productivity of at least 84 percent of the Zambian labor force, employed either in agriculture or in the urban informal sector. Similar circumstances prevail throughout the economies of Sub-Saharan Africa. Efforts to raise labor productivity and wages must build on a systematic effort to dismantle barriers that have limited the pace of structural change in Sub-Saharan Africa to a crawl, even though the experience of Asian economies demonstrates the feasibility of rapidly and substantially transforming poor economies.

China's success has earned it a huge share of the global market for apparel and other labor-intensive products. Steep increases in the cost of Chinese labor and real estate have begun to erode Chinese competitiveness in these sectors. Just as Chinese manufacturers replaced former exporters of labor-intensive goods based in Japan; Taiwan, China; and Hong Kong SAR, China, we now stand on the brink of a new transition that will shift the manufacture of labor-intensive products away from current bases along China's southern and eastern coastline.

This report highlights the opportunity for economies in Sub-Saharan Africa to participate in this transition. Episodes of success within the region, including the export of cut flowers from Ethiopia's newly developed rose plantations and the emergence of Zambia's production of corrugated roofing sheets illustrate the potential inherent in these industries.[13]

The report advocates a package of feasible and inexpensive policies to promote the expansion of labor-intensive light manufacturing. This report, backed by a panoply of comprehensive and detailed research materials, lays out a path that can combine local labor, local (and some imported) materials, and local (and imported) entrepreneurship to jump-start a development process that can expand the capacity of local producers to increase their share in local and eventually overseas markets for garments, leather products, processed agricultural goods, wooden furniture, and simple metal products.

Strategy for a Competitive Light Manufacturing Sector in Sub-Saharan Africa

Sub-Saharan Africa's potential in light manufacturing is huge because of its comparative advantages arising from low labor costs and abundant natural resources. These resources seem well suited for an expansion of manufacturing capacity that can replace imports and capture overseas markets for garments, leather goods, and processed agricultural products.

While exports of wooden furniture and simple metal products may not be feasible in the short term, there is ample local demand for them if they can compete with imports. For example, even though Ethiopia has enormous unexploited potential in timber, particularly bamboo, urban households purchase imported furniture. Why? Because the local price of US$667 a cubic meter of timber in Ethiopia prevents local furniture makers from competing with imports from China, where the corresponding price of timber is US$344, or Vietnam, where it is US$146–US$246. Under these circumstances, policies that support the commercialization of domestic timber resources hold the promise of saving foreign exchange and supporting large numbers of workers at wages well in excess of current earnings in farming or informal jobs.

Africa's potential varies by country and sector, and any strategy to boost the competitiveness of light manufacturing must recognize these specifics. Above all, Sub-Saharan Africa's comparative advantage in light manufacturing rests on the efficiency gains associated with competition in foreign and domestic markets, which guarantees the availability of key inputs at the best quality and lowest cost. Competing imports provide a good benchmark for the sector's global competitiveness.

Sub-Saharan Africa's greatest asset is its large pool of low-wage, less-skilled workers whose productivity in well-managed firms in some subsectors may already approach levels observed in China and Vietnam (tables 1.1 and 1.2). While labor productivity in the average firm is low, with the aid of good management and technical assistance, it can be elevated to move toward the productivity of the best-managed firms. The crux lies in the low skill requirements of light manufacturing subsectors, which make it possible to train, quite cheaply, an unskilled garment worker in about two weeks.[14] While this advantage is unlikely to be permanent, or applicable to all countries, it provides an opportunity to promote new industries that may, like China's, prosper for decades and create millions of productive jobs, much as East Asia did early on and as Bangladesh and Vietnam are doing today.

To move the labor productivity of the average firm toward current domestic best practice does not require providing costly training for all workers. The chief ingredient is new or improved enterprise management together with targeted technical assistance—the impact of which can already be seen in the Ethiopian shoes industry. The detailed studies undertaken as part of this study demonstrate the feasibility of substantially improving management within small firms through inexpensive, short-term training programs (Sonobe, Suzuki, and Otsuka 2011). Furthermore, once begun, the process of upgrading can become self-sustaining, as word spreads of the tangible benefits accruing to early participants in training seminars. Attracting new investors, particularly overseas entrepreneurs who can provide hands-on production and marketing experience in the target industries as well as financial resources and technical expertise, can accelerate the process of industrial expansion and structural change.

Sub-Saharan Africa is fortunate to be well endowed with key inputs and raw materials for light manufacturing. If high-quality inputs are not available locally at the beginning of the expansion process, modest reforms can establish reliable supply chains (for example, for high-quality fabrics or leather) and therefore expand the scope of potentially competitive manufacturing activity. More efficient organization can yield huge savings in transport costs in industries that are located far from a port and require hauling high-volume, high-weight materials. Commercialization of domestic inputs like timber, bamboo, and leather can economize on time as well as foreign exchange and increase the capacity of domestic producers to respond quickly to shifts in demand.

The arithmetic of trade logistics costs validates the enormous potential gains from using Sub-Saharan Africa's input markets to jump-start its light manufacturing product markets. The case of raw materials for Ethiopia's leather products industry illustrates this point. Despite having some of the best hides in the world due to a favorable climate, Ethiopia employs only 8,000 workers in this sector, while Vietnam, which has the same amount of land and population, employs 600,000. As discussed in detail in chapter 8, some short-term policy measures can go a long way to restore Ethiopia's comparative advantage in this sector: eliminate export and import restrictions on leather; implement a program to control "ecto-parasites," a skin disease that adversely affects the quality of leather; and modify bank arrangements to allow the use of cattle as collateral. Longer-term reforms such as modifying land tenure to allow large-scale herding would also help.

The Ingredients of Success

In sum, Sub-Saharan Africa has all the necessary inputs for a competitive light manufacturing sector: a comparative advantage in low-wage labor, abundant natural resources sufficient to offset its low labor productivity relative to its Asian competitors, privileged access to high-income markets for exports, and in most cases a sufficiently large local or regional market to allow emerging producers to develop capabilities in quick-response, high-volume production, and quality control in preparation for breaking into highly competitive export markets. For industries requiring quick access to the coast, Tanzania has a special advantage. But for industries competing with Chinese firms, landlocked Zambia, where there is no such access, instead has natural protection for industries whose products are high in volume and heavy in weight.

We already see instances in which well-managed up-to-scale firms, working in an environment of competition in both product and input markets, have succeeded in raising their productivity to the level of current Asian export leaders. This report proposes that governments in Sub-Saharan Africa follow the course pioneered by a succession of Asian nations by taking the initiative to accelerate the realization of latent comparative advantage in segments of light manufacturing in which specific, feasible, sharply focused, low-cost policy interventions can deliver a quick boost to output, productivity, and perhaps exports, opening the door to expanded entry and growth. In Sub-Saharan Africa, as in Asia, key elements in nurturing developments that can lead to a process of self-sustaining growth include liberalizing access to foreign exchange, imported inputs, and overseas product markets; fostering competition in domestic markets for inputs and outputs; improving domestic infrastructure; offering tax concessions for industrial start-ups; providing short-term training opportunities for workers and managers; and offering generous incentives to attract overseas entrepreneurs capable of filling gaps in domestic stocks of technical knowledge, managerial skills, market experience, and liquid funds.

Growing Markets Inside and Outside Africa

In recent years, four factors have opened new markets for Sub-Saharan Africa's light manufacturing firms.

- First, faster economic growth has expanded the size of the domestic market for manufactures in most countries. New markets thus offer new opportunities.

- Second, foreign investors and aid agencies in Sub-Saharan Africa are investing in the production of manufactures destined for their own markets or foreign markets. Examples include the U.S. Agency for International Development's technical assistance to Rwandese coffee farmers, culminating in a Starbucks contract for exports of locally washed and dried Rwandese coffee to the United States. Similar technical support helped a honey producer to learn how to comply with the phytosanitary standards required to begin exporting fresh honey to Trader Joe's, a U.S. supermarket. The arrival of Chinese, Indian, and Middle Eastern investors is also likely to open doors to more new markets.

- Third, for globally competitive light manufacturing in Sub-Saharan African firms, the market is the world—and that is huge. In 2005 the United States established new trade preferences under the African Growth and Opportunity Act (AGOA), granting Sub-Saharan African products exceptionally favorable access to the United States, while the EU did so under the Cotonou Agreement. These trade preferences are critical to African exporters' success in the global apparel market, for without trade preferences, they are unable to compete with efficient global exporters in the U.S. or EU markets.[15]

- Fourth, regional integration within Africa further increases the attractiveness of growing domestic markets. For example, Ethiopia's participation in regional trade agreements has opened up new markets. Ethiopia is a founding member of two regional blocs: the Intergovernmental Authority on Development (IGAD)[16] and the Common Market for Eastern and Southern Africa (COMESA).[17]

China's Growing Labor Cost Disadvantage: An Opportunity for Africa

Chinese products have penetrated almost every corner of the global market. In 2004 China supplied 18 percent of the total value of the combined U.S. and EU market imports, and in 2008, it supplied 35 percent. To export light manufacturing products, Sub-Saharan producers have no choice but to compete with China, even in the U.S. market.

But the capacity of coastal Chinese firms to outperform rivals based in low-income nations on both price and quality in global markets for labor-intensive light industry manufactures has begun an irreversible process of decline. With the depletion of its large pool of less-skilled workers, rapid cost increases, particularly in wages and nonwage labor expenses, have begun to price growing numbers of China's coastal export firms out of global markets for an expanding array of labor-intensive light industrial products. Some of the displaced production will shift to China's inland provinces, but the rise of new manufacturing clusters is already evident in countries like Vietnam, Cambodia, and Bangladesh, where infrastructure and supply chain arrangements cannot match the enviable circumstances available in coastal Chinese localities like Shenzhen and Dongguan.

Rising wages, stiffening enforcement of labor and environmental regulations, gradual expansion of costly safety net provisions, and the prospect of further increases in the international value of China's renminbi currency mean that the erosion of China's comparative advantage in the export of labor-intensive manufactures will continue and quite possibly accelerate (figure 1.6). This prospect creates an opportunity for Sub-Saharan African nations to jump-start structural changes in their domestic economies that hold the promise of delivering large and sustained increases in output, exports, employment, productivity, and incomes. China's efforts to limit the upward drift of its renminbi currency have

Figure 1.6 Labor Productivity and Average Wage Rates in Manufacturing in China, 1979–2009

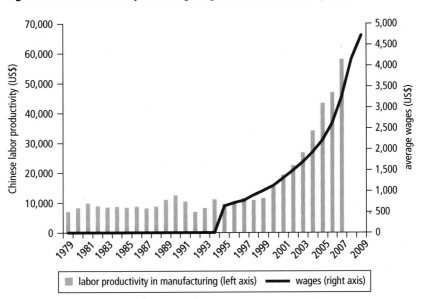

Sources: National Bureau of Statistics of China 2011; Lin 2011.

contributed to substantial domestic inflation, which spurs wage demands and accelerates the erosion of cost advantages in labor-intensive manufactures.

This opens an entry point for other low-wage producers, including firms based in Sub-Saharan Africa, if they can learn to compete. But low-income South Asia, East Asia, and the Middle East and North Africa will be hot contenders for newly available slices of the global market. The challenge facing Sub-Saharan firms is whether they can compete with firms in Bangladesh, India, and Nepal, as well as Vietnam, the Lao People's Democratic Republic, and Cambodia, which have low wages and large pools of less-skilled labor. But even a small slice of the global apparel market would create millions of higher-productivity and higher-wage jobs in Sub-Saharan Africa.

Past Policy Prescriptions: An Intimidating "To Do" List

Is everything a problem? Past studies reviewing constraints to the expansion of light manufacturing in Sub-Saharan Africa often come up with a staggeringly long list, which suggests that no feasible set of policy adjustments could make the region or any country attractive to investors. Most frustrating about this list of constraints is the implication that unless all shortcomings are fixed, the sector cannot grow.

Yet we know that other economies managed to expand production and exports of light manufactures without first resolving the same sorts of constraints currently observed in Sub-Saharan Africa. Visitors to China in the mid-1970s and early 1980s were appalled by low product quality (for example, sewing machines that leaked oil onto the fabric, electric motors that failed in hot, humid weather), passive management (one manager at a large plant insisted that he did not know the unit cost of his product; another, asked to explain the presence of numerous idle workers, said, "If we didn't employ them, where would they go?"), administrative confusion (would-be investors left the Xiamen Special Economic Zone in disgust after managers failed to provide firm prices for land, electricity, or water), delays in moving merchandise through customs and port facilities, lackadaisical attitudes toward customer needs, and others.

Emerging manufacturers in Sub-Saharan Africa must, of course, compete with today's Chinese firms, not with the much weaker Chinese enterprises of the 1980s. But, as noted, powerful market forces have begun to undermine the competitive advantage of China's well-established coastal centers of labor-intensive manufacturing. This process, which has been widely remarked by visitors to the region for at least five years, reflects irreversible forces that seem certain to intensify.

As China's coastal producers of apparel, leather products, and other labor-intensive manufactures suffer a continued squeeze on profitability, they will either shift to other lines of business or move to locations in China's interior, in other Asian nations (Bangladesh, Cambodia, or others), or—as suggested in this report—in Sub-Saharan Africa.

The recommendations of this report, based on intensive study of specific sectors in Ethiopia, Tanzania, and Zambia, draw on the experience of countries like China and Vietnam, which managed to build thriving light industries despite the handicap of problematic initial conditions. The logic underlying our recommendations is simple and direct.

While it will be very difficult for newly emergent African manufacturers to match the price and quality advantages of well-established market leaders in China's coastal regions, the gradual erosion of these firms' competitive advantage has begun and seems certain to continue. This will create growing opportunities initially, for African firms to build up their share of domestic sales of labor-intensive manufactures and, then, with the accumulation of skill, experience, and financial strength, to enter global markets in competition with new entrants from China's interior and from countries like Bangladesh and Cambodia, whose economies suffer from some of the same difficulties and constraints now visible in Ethiopia and its neighbors. As in China and Vietnam, provision of attractive incentives to recruit international entrepreneurs and attract overseas direct investment can accelerate the process and, in particular, expedite the pace of entry into global markets.

The approach proposed in this report is to use intensive study of specific light industry sectors to identify concrete packages of specific, feasible, and inexpensive policy initiatives that can create maximum opportunity to jumpstart the growth of production, employment, and exports.

The specifics of our light manufacturing analysis form the main theme of part II of this report. The following paragraphs focus on economywide issues confronting the nations of Sub-Saharan Africa.

The experience of China, Vietnam, and many other poor nations shows that development prospects benefit from macroeconomic stability and from a business environment that encourages rather than obstructs entrepreneurial initiative. Recent improvements confirm that, for many African countries, the formerly problematic macroeconomic environment is no longer a critical constraint on industrial growth. In fact, Ethiopia is among several nations in Sub-Saharan Africa that have achieved success both in improving macroeconomic conditions (less inflation, lower deficits) and in providing opportunities to private entrepreneurs during the past decade. International experience also shows that manufacturing can expand rapidly, despite the presence of problematic institutional arrangements. In China, for example, laws recognizing the government's obligation to protect the property rights of private businesses appeared only after 2000 and still lack clearly specified implementation mechanisms; in addition, private businesses continue to have little access to bank lending or to formal financial markets.

Although Sub-Saharan Africa receives unfavorable rankings for many constraints identified by the World Bank's Doing Business project, evidence from this study's quantitative and enterprise surveys shows that circumstances facing Sub-Saharan firms are not notably different from conditions surrounding their counterparts in China (Harrison, Lin, and Xu 2011). These constraints may be important, but, as Chinese experience demonstrates, they need not provide impassable barriers to expansion and upgrading.

One effective method of identifying binding constraints that we have pursued in this project is to ask firm managers what they see as the biggest constraints facing their businesses. The results of our investigation include the following.

Although the constraints vary considerably across countries, some patterns appear common (Clarke 2011b). Basic problems such as reliability of power supply, access to finance, access to land, and macroeconomic instability usually dominate at low income levels. Taxes, corruption, and crime become more important as income rises (before falling again). And labor regulation and the availability of skilled workers become more important for middle-income countries (Gelb and others 2007). In addition, we find consistent differences between the perspectives of large and small firms. Managers of small firms express deep concern over limitations arising from their lack of access to finance. Managers of large firms, by contrast, are more likely to identify troublesome labor regulation and inadequate worker education as major issues.

While managers' responses differ across sectors, quantitative analysis typically finds considerable association between measures of firm performance (profits, productivity, growth) and the quality of available infrastructure, particularly infrastructure pertaining to transportation and trade. Some studies have found that burdensome regulation (and labor regulation in particular) is associated with slower growth of firms. By contrast, these studies tend to find little association between firm performance and the prevalence of corruption.

A Selective and Practical Approach: Resolve the Most Critical Constraints in the Most Promising Subsectors

Despite these favorable observations about macroeconomic circumstances and business environment in Sub-Saharan Africa, it is evident that efforts to accelerate the development and structural transformation of African economies confront very substantial obstacles, particularly those that involve finance, infrastructure (electricity, roads), governance, and human capital. Because African governments cannot relax all of these constraints at once, we propose taking a different approach to jump-starting light manufacturing. By focusing on a handful of carefully chosen subsectors, we have been able to leverage value chain analyses and other analytical devices to take a microscopic stock of the constraints in each subsector. Picking reasonable benchmarks and aiming for price competitiveness, we have trimmed the list to a few leading constraints in each subsector. Such priorities make the exercise more manageable, the policy actions more precise, and the sequencing more possible. They are also indispensable in pointing to the few most critical steps that Sub-Saharan governments can take to remove the most serious constraints in the most promising subsectors first and to exploit the potential for light manufacturing.

This study has new features. First, the detailed studies at the subsector and product levels show that constraints vary by country, sector, and firm size. This explains why previous economywide reforms could not relieve the bottlenecks that exist in each sector. The wide range of constraints indicates, first, that the solution to light manufacturing problems cuts across many sectors and does not lie just in manufacturing. Solving the problem of manufacturing inputs requires solving issues in agriculture, education, and infrastructure. Second, precisely because of these links, developing countries cannot afford to wait until all of the problems across sectors are resolved. Instead, a focused approach, such as the one recommended here, is needed. Third, because of the unique structure of Africa's light manufacturing sector, these constraints vary by firm size. Fourth, some of these constraints can be addressed through factory-level measures, others only by government policy, and still others only by strengthening competition.

By identifying and addressing these constraints, African countries can expand light manufacturing production in areas where they have a comparative advantage if they ensure competition in all pertinent markets. Doing so makes the targeted policy solutions practical and feasible within the country's limited financial, fiscal, human resources, and political environment.

Sub-Saharan Africa's potential competitiveness in light manufacturing is based on the following:

- A comparative advantage in low-wage, less-skilled labor. The average monthly wages for a skilled Ethiopian worker in light manufacturing is only 25 percent of that of his/her counterpart in China and 50 percent of that of his/her counterpart in Vietnam. For an unskilled worker, these ratios are 18 percent in China and 45 percent in Vietnam, indicating the significant advantage of an African producer. Expanding these comparisons to include nonwage labor costs, which are high and rising in China but low in Africa, would enlarge the potential for realizing cost savings associated with locating light industry production in African economies.

- An abundance of natural resources, which can supply critical raw materials and inputs such as hides and skins for the footwear industry and abundant land for the commercial cultivation of produce for agribusiness.

While exports of some low-value, heavy-volume products such as furniture or simple metal products may not be feasible, there is enough local demand for them if they are capable of competing with imports. Ethiopia has enormous unexploited potential in timber, particularly bamboo, which makes its furniture industry competitive in the domestic market, creating more productive jobs than in agriculture or the informal sector and saving foreign exchange.

Is there room for Africa in the global market today? China dominates the global export market in light manufactures, and its competitive edge far exceeds that of low-income exporters who recently entered the global market. Fortunately, several global factors can work in Africa's favor if it is able to overcome the key constraints in the most promising subsectors. First, Sub-Saharan Africa has the privilege of enjoying duty-free and quota-free access to the U.S. and EU markets for light manufactures under AGOA and the Cotonou Agreement. Second, rising wages in China's light manufacturing sector present an unprecedented opportunity for Sub-Saharan Africa to take up production of many light manufacturing products and create millions of productive jobs. Given its negligible share in the global market for light manufactures, the potential for growth is huge.

Notes

1. Investment is considerably smaller in Sub-Saharan Africa than in more prosperous developing countries. During 2000–09 Sub-Saharan Africa's investment rose mar ginally from 17 percent of GDP to 19 percent, compared with 26 percent in South Asia, the next richest region. In China and Vietnam, similar to Sub-Saharan Africa in the early 1980s, the share of investment in the economy was 39 and 33 percent, respectively (World Bank 2010).

2. In 2009 per capita income in Sub-Saharan Africa was US$2,051 (purchasing power parity adjusted), only about 36 percent of that in the average developing country (US$5,635).

3. Here we use the broad definition of light manufacturing, which includes the transformation of agricultural products (agribusiness).

4. The data set has "other manufacturing," which is not classified since we do not know its nature.

5. Chinese colleagues described this practice as widespread in 2007; at that time, use of contractors extended to Chinese factories operated by prominent foreign multinational manufacturers of consumer goods (author's field observations).

6. Calculations based on aggregate data for 2007 indicate that average physical labor productivity probably amounted to 1.45 tons per man-day.

7. The policy challenge is to incentivize the average garment firm to mimic well-managed firms. Results of the impact of the Kaizen managerial training programs should confirm whether there is indeed a case for transforming poorly managed into well-managed firms in light manufacturing industries where managerial expertise is a severe constraint. In Ethiopia's case this would apply to all except the leather footwear industry, although here, too, there would be a strong case for sharpening productivity through better management. A firm manager's incentive to upgrade her

skills could be increased by exposing her to greater competition, which would weed out inefficient firms.

8. Clarke (2011a) also finds that, although labor costs in many African countries, particularly in the resource-based economies, appear to be higher relative to those in East Asia countries that have been relatively successful in manufacturing, many still compare favorably with East Asian countries, after both labor costs and labor productivity are considered.

9. Here the business environment is broadly defined to include crime, political stability, and government expropriation including corruption.

10. The share of manufactures in Chinese exports: 49.7 percent in 1980; 41.6 in 1970; 42.8 percent in 1959. Data for 1980 are from National Bureau of Statistics of China (2005, 68–69); data for 1959 and 1950 are from Eckstein (1977, 252).

11. Firm growth is measured as the average change in log sales over 2008–10. The top and bottom 5 percent of the observations are dropped to ensure robustness to outliers.

12. Although it is difficult to compare the size of the informal sector across countries due to difficulties with both definitions and measurement, most evidence suggests that the informal sector is larger in Sub-Saharan African than in other regions. Schneider (2005) estimates that the informal sector accounted for about 41 percent of GDP in the 24 African countries with data. This is similar to its share in Latin America, but higher than in most other regions.

13. In Zambia, the zero-rating of rolled steel imports in 2000 enabled the emergence of some 20 firms to begin producing corrugated roofing sheets. Zambia experienced a shift from trading in imported roofing sheets toward manufacturing them itself and even exporting them to neighboring countries.

14. According to survey results for light industries, "In Ethiopia and China, 85 percent to 90 percent [of firms] report that it takes at most four weeks for new workers to be fully trained" (Fafchamps and Quinn 2011; Sonobe, Suzuki, and Otsuka 2011, 16).

15. Conway and Shah (2010). The recent decimation of Madagascar's apparel production after the removal of its AGOA eligibility is a case in point.

16. Other members of IGAD include Djibouti, Ethiopia, Eritrea, Sudan, Somalia, Uganda, and Kenya.

17. The COMESA was established in November 1993 in Kampala, Uganda. It has 20 member states that stretch from the Arab Republic of Egypt in the north to Swaziland in the south. The current members are Angola, Burundi, the Comoros, the Democratic Republic of Congo, Djibouti, Egypt, Eritrea, Ethiopia, Kenya, Madagascar, Malawi, Mauritius, Namibia, Rwanda, the Seychelles, Sudan, Swaziland, Uganda, Zambia, and Zimbabwe (Tanzania recently left COMESA). Before the formation of COMESA in 1993, the regional community was known as the Preferential Trade Area for Eastern and Southern Africa (PTA), which was established in September 1981 and had a different treaty than the COMESA. All previous PTA members, except the Comoros and Somalia, have signed the COMESA treaty. COMESA has a combined population of close to US$400 million and GDP of about US$170 billion, respectively. The total surface area is more than 9 million square kilometers, of which 60 percent is endowed with rivers and lakes with a potential for irrigation,

fisheries, and hydroelectric power generation. Less than 10 percent of the arable land in the region is under cultivation and only 5 percent of available water is used for cultivation. The region has used only 4 percent of its hydroelectric potential. The region is also a source of wealth of minerals (Birega 2004, 14).

References

Amoako, K. Y. 2011. "The Africa Transformation Report." Presentation at the African Center for Economic Transformation "Workshop on Growth and Transformation in Africa," Bellagio, Italy.

Arbache, Jorge, Delfin S. Go, and John Page. 2008. "Is Africa's Economy at a Turning Point?" in *Africa at a Turning Point? Growth, Aid, and External Shocks*, ed. Delfin Go and John Page, 13–85. Washington, DC: World Bank.

Banister, Judith. 2005. "Manufacturing Earnings and Compensation in China." *Monthly Labor Review* (August): 22–40.

Birega, Gebremedhine. 2004. "Preliminary Country Paper of Ethiopia on Competition Regime: Capacity Building on Competition Policy in Select Countries of Eastern and Southern Africa." AHa Ethiopian Consumer Protection Agency, Addis Ababa.

Chenery, Hollis. 1980. "Interactions between Industrialization and Exports." *American Economic Review* 70 (2): 281–87.

Clarke, George. 2011a. "Assessing How the Investment Climate Affects Firm Performance in Africa: Evidence from the World Bank's Enterprise Surveys." Background paper (Light Manufacturing in Africa Study). Available online in Volume III at http://econ.worldbank.org/africamanufacturing. World Bank, Washington, DC.

———. 2011b. "Wages and Productivity in Manufacturing in Africa: Some Stylized Facts." Background paper (Light Manufacturing in Africa Study). Available online in Volume III at http://econ.worldbank.org/africamanufacturing. World Bank, Washington, DC.

Conway, Patrick, and Manju Shah. 2010. "Incentives, Exports, and International Competitiveness in Sub-Saharan Africa: Lessons from the Apparel Industry." World Bank, Washington, DC.

Eckstein, Alexander. 1977. *China's Economic Revolution.* Cambridge, U.K.: Cambridge University Press.

Fafchamps, Marcel, and Simon Quinn. 2011. "Results from the Quantitative Firm Survey." Background paper (Light Manufacturing in Africa Study). Available online in Volume III at http://econ.worldbank.org/africamanufacturing. World Bank, Washington, DC.

Fox, Louise. 2011. "Why Is the Informal Normal in Low-Income Sub-Saharan Africa?" Presentation at the World Bank, Washington, DC.

Gelb, Alan, Vijaya Ramachandran, Manju Kedia Shah, and Ginger Turner. 2007. "What Matters to African Firms? The Relevance of Perceptions Data." Policy Research Working Paper 4446, World Bank, Washington, DC.

Harrison, Ann E., Justin Y. Lin, and L. C. Xu. 2011. "Explaining Africa's (Dis) Advantage." Background paper (Light Manufacturing in Africa Study). Available online in

Volume III at http://econ.worldbank.org/africamanufacturing. World Bank, Washington, DC.

Harrison, Ann, and Andres Rodríguez-Clare. 2010. "Trade, Foreign Investment, and Industrial Policy for Developing Countries." In *Handbook of Development Economics*, Vol. 5 of Dani Rodrik and Mark Rosenzweig eds., 4039–214. Amsterdam: North-Holland.

Huang, Yiping. 2010. "China's Great Ascendancy and Structural Risks." *Asian-Pacific Economic Literature* 24 (1): 65–85.

Lin, Justin. 2010. "New Structural Economics: A Framework for Rethinking Development." Policy Research Working Paper 5197, World Bank, Washington, DC.

———. 2011. "From Flying Geese to Leading Dragons: New Opportunities and Strategies for Structural Transformation in Developing Countries." Policy Research Working Paper 5702, World Bank, Washington, DC.

McMillan, Margaret, and Dani Rodrik. 2011. "Globalization, Structural Change, and Productivity Growth." International Labour Organisation, World Trade Organization, Boston, MA.

National Bureau of Statistics of China. 2005. *China Compendium of Statistics 1949–2004*. Beijing: China Statistics Press.

———. 2011. *China Statistical Yearbook 2010*. Beijing: China Statistics Press.

Panagariya, Arvind. 2008. *India: The Emerging Giant*. London: Oxford University Press.

Rankin, Neil, Justin Sandefur, and Francis Teal. 2010. "Learning and Earning in Africa: Where Are the Returns to Education High?" CSAE WPS/2010-02, Centre for the Study of African Economies, Department of Economics, Oxford University, Oxford, U.K.

Schneider, Friedrich. 2005. "Shadow Economies around the World: What Do We Really Know?" *European Journal of Political Economy* 21 (3): 598–642.

Söderbohm, Måns. 2011. "Firm Size and Structural Change: A Case Study of Ethiopia." Paper prepared for the Plenary Session of the African Economic Research Consortium "Biannual Research Workshop," Nairobi, Kenya, May.

Sonobe, Tetsushi, Aya Suzuki and Keijiro Otsuka." 2011. "Kaizen for Managerial Skills Improvement in Small and Medium Enterprises: An Impact Evaluation Study". August. Background paper prepared for the Light Manufacturing in Africa Study, available online as Volume IV at http://econ.worldbank.org/africamanufacturing.

Sutton, John, and Nebil Kellow. 2010. *An Enterprise Map of Ethiopia*. London: International Growth Center.

UNCTAD (United Nations Conference on Trade and Development). 2011. "Best Practices in Investment for Development: Case Studies in FDI; How to Integrate FDI and Skill Development, Lessons from Canada and Singapore." Investment Advisory Series, Series B, no. 5. United Nations, New York.

World Bank. 2009. "Ethiopia: Toward the Competitive Frontier; Strategies for Improving Ethiopia's Investment Climate." Report 48472 ET, Finance and Private Sector Development, Africa Region, World Bank, Washington, DC.

———. 2010. *World Development Indicators 2010.* Washington, DC: World Bank.

———. 2011. *More Jobs and Prosperity in Zambia: What Would it Take?* Report 62376-ZM, Finance and Private Sector Development Unit, Africa Region, World Bank, Washington DC.

World Economic Forum. 2011. *The Africa Competitiveness Report 2011.* Washington, DC: World Bank; Geneva: World Economic Forum.

Zhao, Shinning, and Jie Zhang. 2010. "Impact of Employment Contracts Law on Employment Relations in China." *Indian Journal of Industrial Relations* 45 (4, April).

Part 2

What Constrains Light Manufacturing in Sub-Saharan Africa?

The World Bank's Regional Program for Enterprise Development has conducted a comprehensive set of enterprise surveys that provide perception-based and objective data. Previous analytical work to identify the investment climate constraints to firm growth in developed and developing countries has followed two broad approaches: conducting perceptions-based surveys of firm managers, asking what they see as the biggest barriers to their firm's operations and growth; and correlating firm performance and various objective and subjective measures of the investment climate. Both approaches typically lead to a long list of constraints, including electricity, corruption, crime, inadequately educated workforce, labor regulations, business licensing and permits, and so on.

But African countries, given their limited resources, cannot afford to wait until all of the problems across all sectors and locations have been resolved. Instead, they should focus on reducing the constraints in sectors demonstrating good potential for competitiveness and employment growth. This report highlights light manufacturing as a promising starting point. The first step is to identify the most critical constraints for each country, for each light manufacturing subsector, and for firms of different sizes. Recent studies use rigorous analytical methods to evaluate owners' perceptions and firm-specific information from the enterprise surveys to identify the constraints on firm growth. This approach was also applied to research papers commissioned for this study to narrow the possible list of constraints (Harrison, Lin, and Xu 2011; Dinh, Mavridis, and Nguyen 2010).

In a review of the empirical literature on the investment climate and its effects on firm performance in Africa, Clarke (2011) finds that more than two-thirds of firm managers identify access to finance and electricity as their top concerns in more than two-thirds of African countries. Access to finance tends to be a greater concern for managers of small firms, and labor regulation and worker education are greater concerns for large firms.

Main Constraints in Ethiopia, Tanzania, and Zambia

The findings in this part are drawn from the following sources:

- A literature review of empirical work on the investment climate facing firms in Africa as well as research papers aimed at narrowing the possible list of constraints

- Qualitative interviews of about 300 enterprises (both formal and informal of all sizes) by the study team in three African countries (Ethiopia, Tanzania, and Zambia) and two Asian countries (China and Vietnam), based on a questionnaire designed by Professor John Sutton of the London School of Economics[1]

- Quantitative interviews of about 1,400 enterprises (both formal and informal of all sizes) by the Centre for the Study of African Economies at Oxford University in the same countries, based on a questionnaire designed by Professor Marcel Fafchamps and Dr. Simon Quinn of Oxford University

- Comparative value chain and feasibility analysis based on in-depth interviews of about 300 formal medium enterprises in the same five countries, conducted by the consulting firm Global Development Solutions, Inc.

- A Kaizen study on the impact of managerial training for owners of small and medium enterprises in Ethiopia, Tanzania, and Vietnam.

These materials provide complementary perspectives. The literature review allows us to build on previous findings. The comparative value chain analysis, based on a small sample of relatively large firms, allows us to benchmark the cost and productivity of African firms with those of their Asian competitors. The quantitative surveys have a much larger sample and focus on smaller firms

Figure II.1 Sources of Excess Production Costs of Medium Firms in Africa: Average across Ethiopia, Tanzania, and Zambia as a Percentage of Chinese Production Cost

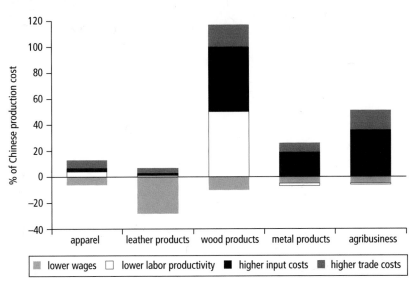

Source: Global Development Solutions 2011.

(including informal ones) but do not measure cost and productivity. The qualitative interviews, which cover both small and larger firms, provide insight on issues that the other studies overlook, especially access to industrial land. And the Kaizen study shows how some policy interventions can help owners and entrepreneurs to improve firm performance.

In four of five subsectors, Africa's advantage in money wages is wiped out by higher input and trade logistics costs (figure II.1). Leather products are an exception, with the highest share of labor in production costs (45 percent in China as opposed to around 20 percent in the others). Among the up-to-scale and fairly well-managed companies, the production cost penalty due to lower labor productivity is small (except in wood products), is shielded from international competition by high transport costs, and benefits little from technical assistance for firms and workers.

Together, the analytical tools and sources of information used for this study indicate that in Ethiopia, Tanzania, and Zambia and across subsectors and sizes, six main constraints impede the competitiveness of light manufacturing: the availability, costs, and quality of inputs; access to industrial land; access to finance; lack of entrepreneurial skills, both technical and managerial; lack of worker skills; and poor trade logistics. For small firms, entrepreneurial skills, land, inputs, and finance are the most important constraints, while for large firms, trade logistics, land, and inputs are among the more important. The six chapters in this part address each of them in turn.

Note

1. This questionnaire and all the others mentioned in this paragraph can be found online at http://econ.worldbank.org/africamanufacturing.

References

Clarke, George. 2011. "Assessing How the Investment Climate Affects Firm Performance in Africa: Evidence from the World Bank's Enterprise Surveys." Background paper (Light Manufacturing in Africa Study). Available online in Volume III at http://econ.worldbank.org/africamanufacturing. World Bank, Washington, DC.

Dinh, Hinh T., Dimitris Mavridis, and Hoa B. Nguyen. 2010. "The Binding Constraint on Firms' Growth in Developing Countries." Background paper (Light Manufacturing in Africa Study). Available online in Volume III at http://econ.worldbank.org/africamanufacturing. World Bank, Washington, DC.

Global Development Solutions. 2011. "The Value Chain and Feasibility Analysis; Domestic Resource Cost Analysis." Background paper (Light Manufacturing in Africa Study). Available online as Volume II at http://worldbank.org/africamanufacturing. World Bank, Washington, DC.

Harrison, Ann E., Justin Y. Lin, and L. C. Xu. 2011. "Explaining Africa's (Dis) Advantage." Background paper (Light Manufacturing in Africa Study). Available online in Volume III at http://econ.worldbank.org/africamanufacturing. World Bank, Washington, DC.

Chapter 2

Input Industries

Large and small firms alike identify input supply—including availability, quality, and cost—as a leading obstacle to developing competitive light manufacturing in the three African countries that are the focus of this report. Inputs are a binding constraint in two of five light manufacturing subsectors (agribusiness and wood products) and an important constraint in the other three (apparel, leather, and metal products). On average across the five subsectors, inputs (which represent more than 70 percent of total production costs) are more than 25 percent more expensive in Africa than in China (figure 2.1), a 20 percent production cost penalty. In most cases higher input costs wipe out Africa's labor cost advantage.

Since farm products and wood are major inputs for four of the five light manufacturing sectors that this report identifies as particularly suitable for policy attention, improving the performance of agriculture and forestry quickly emerges as a key item on the policy agenda for enhancing the competitiveness of African light manufacturing.

The main input policy issues are import tariffs (all sectors), price controls and export bans on agricultural products, barriers to the import and distribution of high-yield seeds, difficulties in obtaining access to land and finance for commercial farming, livestock, and forestry, and disease control in the livestock sector.

Input issues are compounded by poor trade logistics, which add to the costs and delays of importing inputs that cannot be sourced locally. Currency overvaluation and instability also compound the input issues.

Effects of Input Costs on Competitiveness

Without the ability to acquire large volumes of diverse inputs at competitive prices, of consistent high quality, and on short notice, Africa-based firms cannot hope to achieve competitiveness in international markets. With inputs accounting for more than 70 percent of the total cost for light manufacturing products, a small variation in the price paid for inputs can wipe out any labor cost advan-

Figure 2.1 Impact of Higher Input Costs on Total Production Costs in Relation to China

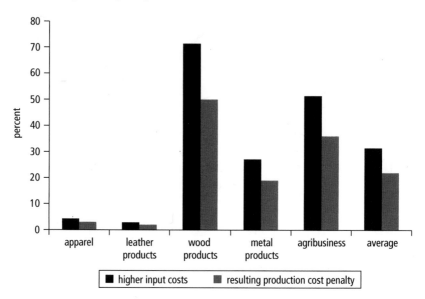

Source: Global Development Solutions 2011.

tage a country may have (figure 2.1). And price competitiveness alone is not enough. Countries increasingly compete on the basis of their capacity to deliver on time large quantities of products with consistently high quality. This requires access to diversified, reliable, and plentiful sources of quality inputs. Even simple products require many inputs—a padlock, for example, has 12 parts.

Companies in China's coastal export belt enjoy favorable prices for many inputs, immense choice, and the ability to source large quantities at consistently high quality with quick delivery times. The reason: China has achieved economies of scale and scope through its industrial clusters. Many of them were built around the sources of raw materials, such as the wood cluster in Nankang and the sugar cluster outside Ganzhou. China's coastal regions also have extremely competitive input industries—for example, in textiles and steel. A typical Chinese garment manufacturer in Shanghai or Dongguan can choose among thousands of domestic fabric suppliers, many of them within a short distance in a textile or garment cluster, all competing to sell fabric, buttons, ribbon, zippers, or other supplies. Also relevant is the contribution of official Chinese efforts, often at the local level, to facilitate the development of input (and output) markets as a quasi-public good: the Yiwu commodity market in Zhejiang Province, built through local government initiative, is now the largest in the world (Ding 2010).

Moreover, reforms and investments in agriculture and forestry have enabled both China and Vietnam to develop competitive input industries for agribusiness and wood products. Both can import the inputs they need—for instance, high-end textiles from Japan and the Republic of Korea, high-end sheep and goat leather from Australia, and pine wood from the Russian Federation—at low cost and at fast turnaround times thanks to liberal import policies, world-class transport, and vastly improved trade logistics.

Why Are Input Costs Higher?

Agricultural and forestry policies make wheat, animal feed, and (legal) wood—all of which could be produced competitively in Africa—40 percent more expensive than in China. Domestic livestock policies also constrain the volume of quality leather, which could be competitively produced. High import tariffs and transport costs make steel 30 percent more expensive than in China. The absence of a competitive domestic textile industry imposes a cost penalty that could be manageable with good trade logistics (as in Vietnam, where having to import textiles adds a 3 percent production cost penalty). Firms aiming to supply domestic markets—a category that includes most small firms in Africa—are further penalized by having to pay duties on imported inputs.

Agribusiness

The price of wheat hovers around US$200–US$250 per ton in Asia, compared with US$300–US$350 in Ethiopia and Tanzania and over US$400 in Zambia (table 2.1). Since wheat accounts for more than 80 percent of the cost of flour, the higher prices of wheat impose a production cost penalty of at least 40 percent.

Table 2.1 Wheat Prices in the Five Countries, 2010 Average

Price indicator	China	Vietnam	Ethiopia	Tanzania	Zambia
Raw material inputs as % of value chain	85	81	93	86	87
Total cost of raw material per ton of wheat flour (US$)[a]	322	323	408	363	625
Domestic wheat					
Cost per ton (US$)	192	269	333	300–365	400–500
% of total inputs	60	73.7	78.1	—	68
Imported wheat					
Cost per ton (US$)	None	208	304	261–328	None
% of total inputs	None	—	—	—	None

Source: Global Development Solutions 2011.
Note: — = Not available.
a. Amount reflects by-product content in wheat grains, conversion losses, and waste. Includes 10% duty for imported wheat in Tanzania.

A ton of wheat costs US$330 in Ethiopia, compared with about US$200 in China and Vietnam. With Ethiopian wheat yields generally less than 1 ton per hectare, versus 4 tons in Vietnam and up to 6 tons in China, Africa faces great challenges in producing staple crops:

- Ethiopia and Tanzania have 10 percent import tariffs on wheat. Tanzania also has a 60 percent import tariff on wheat flour (import tariffs in Vietnam are only 2 percent). China has 15 percent import tariffs on wheat, but produces it at the world price (table 2.1).
- Importing, producing, and distributing high-yielding seeds and fertilizers are limited by policy.
- Land policy discourages irrigation and the entry of investors aiming to work with smallholders.
- Storage infrastructure remains poor.
- Inadequate market mechanisms fail to deliver stable, predictable prices.
- Wholesalers lack working capital.

Leather Products

Despite abundant livestock and skins in Africa, firms have difficulty finding large volumes of quality African leather because of the lack of veterinary services (diseases reduce the quality of skins), the lack of access to finance for small herd holders (who cannot use their land or animals as collateral), and difficult access to land for potential investors in large herds as a result of the challenging combination of traditional and modern land rights. Measures to discourage the export of leather reduce the profitability of tanneries and thus impede the development of a competitive supply chain. The tannery business is also capital and energy intensive, so facilitating imports of quality leather may be a more viable alternative to accelerate the manufacture of leather products in the short term.

Leather products illustrate how limited, inexpensive intervention can activate Ethiopia's latent export potential. A U.S. Agency for International Development study finds that the infestation rate of ectoparasites, which cause a skin disease in animals in Ethiopia, could be substantially reduced from 90 to 5 percent with four treatments a year for each animal, costing about US$0.10 for all four treatments (USAID 2008). The total cost for such a program covering the whole country would be less than US$10 million a year, a modest amount in relation to the potential benefits. Our investigations show that allowing processed leather imports and eliminating import duties on leather and on other shoe parts (chemicals, glues, treads, laces, and soles) would enable Ethiopian shoes and other leather goods to become competitive in international markets in advance of any improvement in the currently poor state of trade logistics.

Importing leather would incur a US$1 cost penalty per pair of shoes (a 6 percent production cost penalty), which would be more than offset by Ethiopia's US$5 labor cost advantage per pair of shoes (US$3.50 advantage for Tanzania and Zambia). Removal of import duties on leather and other shoe parts would also expand opportunities for small-scale production of shoes and other leather goods for domestic consumption.

Apparel
The situation facing apparel manufacturers resembles conditions in the leather sector. Raw materials (fabrics, collars, threads, buttons, and the like) account for more than 70 percent of polo shirt production costs. The inability of African producers to supply most inputs for export-quality shirts would be manageable with good trade logistics, as material imports add only a 3 percent production cost penalty. But poor trade logistics almost double this penalty by adding a further 5 percent to the price of imported inputs. As in the case of leather, the Ethiopian government's policy of encouraging investments in textile mills by banning the export of cotton may backfire by discouraging cotton production. Again, nonexporters are further penalized by having to pay import duties on their inputs. This policy even penalizes exporters, which cannot leverage small firms and sell their material waste to them. As with leather, removing tariffs on imported inputs (fabrics, buttons, zippers, ribbons) could enhance the prospects for manufacturers targeting both the domestic and export markets and might also stimulate domestic production of upstream products—in this case, fabrics and cotton.

Wood Products
Although Africa is a major global source of precious wood, few countries have developed (legal) competitive sources of nonprecious wood for their domestic economies. They should do so for economic reasons (rapidly growing domestic demand as a result of population growth, urbanization, and income growth) and environmental reasons (most of the wood for domestic use is from illegal logging). The price of legal nonprecious wood is significantly higher in the three African countries than in China and Vietnam (table 2.2). Since lumber accounts for more than 70 percent of the cost of simple wood products (such as wooden chairs and doors), this price differential burdens Ethiopian makers of wood products with production cost penalties (versus imported goods) that average as much as 35 percent.

Legal wood is more expensive in Africa because of the absence of sustainable industrial scale plantations of fast-growing species (which could be developed on degraded land close to urban centers and ports) and the levy of high taxes (in Cameroon taxes increase the price of legal wood by 40 percent).

Table 2.2 Pine Lumber Prices in the Five Countries

Country	Price per cubic meter (US$)
Ethiopia	667
Tanzania	275
Zambia	394
China	344
Vietnam	146–246

Source: Global Development Solutions 2011.

Metal Products

Tin-free steel is US$1,610 a ton in Zambia, US$1,414 in Ethiopia, and US$1,106 in China, imposing a production cost penalty in excess of 20 percent on makers of simple metal products, for which steel is more than half the cost. Half the price difference comes from import tariffs on the steel destined for domestic consumption (10 percent in Ethiopia and 20 percent in Zambia). The other half comes from transport costs for imported steel, a low value-to-weight item.

Possible Solutions from Asia

Africa has comparative advantages in many subsectors of agriculture, livestock, and forestry. This report recommends that governments take these strengths as the foundation for policy initiatives aimed at kick-starting light manufacturing industries that make use of these potentially abundant inputs to create much-needed nonfarm employment, expand the supply of foreign exchange through import replacement and export expansion, and also increase the demand for crops, animal products, timber, and other low-cost sectors with ample scope for future production increases. Such initiatives hold the promise of sparking an upward spiral of growth, structural transformation, productivity, employment, and penetration of global markets.

To implement this strategy, African nations must remove obstacles to the expansion of light manufacturing. As we have seen, the chief difficulty involves insufficient access to readily available, reasonably priced materials like hides and fabrics of suitable quality.

China and Vietnam faced similar difficulties at the start of their accelerated entry into global markets for labor-intensive manufactured goods. They responded by pursuing a two-track strategy to facilitate access to inputs. First, they developed, early on, world-class trade logistics to support imports of the inputs that could not be competitively sourced domestically (chapter 5). Second, they reformed and provided support to help key input industries to become competitive.

China also transferred large numbers of government-owned enterprises to partial or full private ownership, instituted reforms aimed at commercializing the remaining state enterprises, dismantled many official controls over prices and resource allocation, encouraged foreign direct investment for key inputs (as in machine manufacturers), and developed sustainably managed wood plantations and competitive agricultural sectors. National and especially local governments supported the creation of input and output markets through the provision of land and financing (as for the Yiwu market in China's Zhejiang Province, developed with the initiative and support of local officials) and coordination along the value chains. China exempted exporters from taxes and duties on imported materials. It also provided information and technical assistance on inputs, technology, and suppliers to small and medium enterprises.

Policy Recommendations

To address input costs and quality, Sub-Saharan governments should adopt a two-prong strategy: they should facilitate access to inputs for light manufacturing by working to improve trade logistics and pushing to deepen regional integration (chapter 5); at the same time, they should promote efforts to develop the potentially competitive industries that supply key inputs for light manufacturing.

- *Remove import tariffs on all inputs for light manufacturing, even for products destined for national and regional markets.* Preferences are now given to exports through tariff exemptions on imported inputs for exported products. Such exemptions create a significant disincentive for manufacturers to produce for both the domestic and export markets, restricting their ability to maximize profits in both markets. They also divide exporters (mostly medium and large firms) and nonexporters (mostly small and micro firms), hurting both (large firms cannot subcontract with small firms, as they do in China). Removing import tariffs is a low-cost reform: for Ethiopia, eliminating tariffs on leather, fabric, and other materials used by light manufacturers would reduce fiscal revenues by only 2 percent (chapter 8).

- *Remove restrictions on exports of key light industry inputs such as unprocessed leather and cotton.* Export bans are well intentioned but self-defeating. The objective is to secure a sufficient and cheap supply of such raw materials to encourage domestic processing of goods like cotton and leather. But prohibiting exports eliminates an important component of demand for these products. This may cause an unintended chain of damaging consequences: a reduction in the output of the targeted materials, an increase in their domestic prices, erosion of potential cost advantage for manufactures that rely on these materials, and a reduction (rather than the intended increase) in domestic processing of these materials. While freeing up exports of cotton and leather will increase demand, and hence encourage their production, allowing imports of these commodities will put competitive pressure on domestic producers to improve quality and give manufacturers the flexibility to source the cheapest and best-quality inputs, thus stimulating employment growth and structural transformation.

- *Liberalize input markets in agriculture.* That is, remove the barriers to importing and distributing high-performing planting materials. This is often the most important critical step in developing a competitive agriculture sector. Indeed, there is little point in investing in fertilizers, skills, and irrigation if high-yielding seeds are not available.

- *Facilitate access to finance for smallholders in agriculture, livestock, and forestry.* Smallholders should be able to use land, livestock, and agricultural output as collateral.

- *Facilitate access to land for strategic investors in the agriculture, livestock, and sustainable forestry sectors.* Access should be facilitated through an inclusive and transparent process that brings clear benefits to local communities and mitigates environmental impacts. Good-practice new entrants (such as strategic first movers) can do much to improve the performance of key input industries. The reforms discussed above would be essential to attracting such investors, but they may be long in coming, especially the politically and socially sensitive land reforms. So, African governments may want to remove some of the key constraints preventing the entry of potential strategic first movers. The Ethiopian government did this by giving 7 hectares to the first rose farm, creating the second largest export industry and, with it, 50,000 new jobs.

- *Provide technical assistance to smallholders.* Technical assistance could help smallholders to connect with strategic investors in agriculture, livestock, and forestry through contract farming around nucleus farms and plantations and also encourage them to respond to new market opportunities. For example, if veterinary extension services would teach smallholders to protect cattle from pests, farmers would be able to supply hides of acceptable quality to tanneries serving emergent domestic manufacturers of leather goods.

- *Provide public goods in agriculture, livestock, and forestry.* Such public goods include imposing disease controls, promoting standards, fighting illegal logging, and reducing taxes on legal wood.

This chapter has discussed some issues related to input industries and proposed some general solutions. But further in-depth technical analysis of these input industries will be needed for each individual country to solve specific industrial problems, such as how to facilitate the entry of competitive producers in livestock (leather and milk), food staples (wheat), industrial crops (cotton), and sustainable forestry (wood) in a way that benefits local communities and preserves the environment. From a political economy perspective, existing distortions and restrictions in input industries inevitably provide benefits to some stakeholders even as they impose much larger costs on the economy by impeding the growth of manufacturing and, with it, the expansion of employment, productivity, and exports. In proposing and implementing reforms, analysis will need to identify the losers and demonstrate that the private losses arising from reforms will be dwarfed by the large-scale, economywide benefits to be reaped following successful implementation of the reforms proposed in this report.

References

Ding, Ke. 2010. "The Role of the Specialized Markets in Upgrading Industrial Clusters in China." In *From Agglomeration to Innovation: Upgrading Industrial Clusters in Emerging Economies*, ed. Akifumi Kuchiki and Masatsugu Tsuji, 270–89. Houndmills, U.K.: Palgrave Macmillan.

Global Development Solutions. 2011. "The Value Chain and Feasibility Analysis; Domestic Resource Cost Analysis." Background paper (Light Manufacturing in Africa Study). Available online as Volume II at http://worldbank.org/africamanufacturing. World Bank, Washington, DC.

USAID (U.S. Agency for International Development). 2008. "Success Story: Ethiopians Learning to Fight Ectoparasites." Financial Transactions and Reports Analysis, USAID, Washington, DC.

Chapter 3

Industrial Land

Lack of access to industrial land can cripple efforts by both smaller and larger firms to take advantage of market opportunities and attain a competitive operational scale. Smaller firms need land to set up and expand a business, larger firms need it to expand their factories, and both can benefit from using land as collateral to obtain loans.

It is ironic that land is a constraint for most manufacturing firms in land-abundant Sub-Saharan Africa. As all manufacturing firms need industrial land, equipped with utilities and transport links to markets, Sub-Saharan Africa's huge deficit in industrial land puts land policy at the core of its industrial development agenda (Limão and Venables 2001; Dollar, Hallward-Driemeier, and Mengistae 2004; Hausman, Lee and Subramanian 2005).

In the early 1980s similar problems also emerged in China and Vietnam. But efficient government facilitation of access to industrial land, first in special economic zones and then throughout the domestic economy, resolved the land constraint in China and made them less binding in Vietnam.[1]

Effects of Industrial Land on Competitiveness

This section discusses what kind of land firms need, how land policy shapes the existing pattern of land use, and why the availability of affordable and accessible land is vital for the growth of light manufacturing.

Affordable Land, Reliable Utilities, and Accessible Locations

Qualitative interviews suggest that problems acquiring land often prevent firms in Sub-Saharan Africa with 4–5 employees from growing into businesses with more than 10–15 employees. To do so, they would need a larger workspace connected to affordable and reliable utilities and offering reliable transport links to markets for inputs and outputs. Most small firms are located in the owner's home or in small workshops. Connecting to utilities requires large fixed investments that are typically beyond the means of small informal operators.

Small business owners typically have insufficient savings and retained earnings with which to purchase industrial land to expand the business. Where the government tries to provide factory shells outside cities, transport costs for hauling raw materials and products are an additional constraint. In a rare model where government allocates small plots of industrial land to small cooperatives, the rules of the cooperatives make it difficult for small firms to grow.

Chinese governments, especially at the local level and in coastal provinces, have gradually developed policies (initially inspired by the early success of a few special economic zones) that enable smaller firms to expand either in organic (naturally developed via market forces) or synthetic (set up by authorities) industrial clusters or in industrial parks that jointly ease land and infrastructure constraints. For firms employing 35–50 persons and looking to expand production, the local governments are increasingly able to facilitate entry into industrial parks that provide "plug-and-play" factory shells to ambitious entrepreneurs.

Problems of access to industrial land diminish for larger firms, although large firms in Sub-Saharan Africa have much more difficulty accessing land than firms in similar size categories in East Asia. In Ethiopia, exporters (typically the larger firms) in the preferred sectors (apparel and leather products) receive preferential access to cheap industrial land. More generally in Sub-Saharan Africa, industrial zones are usually reserved for large exporters, most often firms with foreign ownership (Farole 2011).

Access to Larger Plots of Land and Economies of Scale

The contribution of industrial agglomeration to the growth of firm productivity has been confirmed for the developed countries (Ciccone and Hall 1993; Lall 2004; Lee and Zang 1998), but similar effects are evident in Chinese clusters as well (Fan and Scott 2003; He and Zhu 2007; Pan and Zhang 2002). Industrial agglomerations in China and Vietnam enable firms to avoid problems related to inadequate land markets and difficulty purchasing large plots of land by locating them in industrial parks where they can reap scale economies.

Clustering in China generated large spillovers, most visible in liberalized and export-oriented industries in the 1980s. And by the 1990s, the spillovers from clustering were visible in nearly all industries, confirming the benefits of China's industrial clustering strategy, fashioned after that of the Republic of Korea (Lee and Zang 1998). The strategy had clear payoffs for the labor-intensive and globalized industries, particularly in coastal regions such as Zhejiang, Jiangsu, and Guangdong provinces (He and Wang 2010). Clustering, often as a direct result of local government initiatives—for example, the effort of county leaders in Xinji Municipality, Hebei Province, to develop a local specialty in leather processing and manufacturing—also contributed more broadly to China's growing industrial competitiveness (on Xinji, see Blecher and Shue 2001; Wang 2001; He, Wei, and Xie 2008).

The Chinese industrial parks with access to plug-and-play factory shells enable firms to scale up to hundreds or thousands of employees in a few years. By incurring the fixed costs of utility-equipped factory shells, the government precluded the firms' need for finance to construct the factories. In recent years, Chinese industrial parks have gradually created one-stop shops that allow firms located under their jurisdiction to simplify and speed up official registration and regulatory transactions. Western buyers have an appetite for large orders, and China's industrial parks allow firms to scale up rapidly to meet the buyers' needs. Such arrangements reduce search and transaction costs and make information easily available to buyers and sellers.

In Vietnam, the government has a very location- and industry-specific industrial land policy that actively courts large manufacturing firms—usually foreign direct investment–financed or state-owned enterprises—and gives them large facilities in well-designed industrial parks, much like in China. Although this strategy has fueled Vietnam's export-led growth, success is mixed. The government seems to have overinvested in about 233 parks, which have an average occupancy rate of 50 percent. The infrastructure investments to increase the supply of industrial land have been very successful in parks near cities but not in remote or backward areas,[2] where the objective is to reduce poverty.[3] The industrial parks have worked especially well for larger foreign firms. Domestic investors, especially smaller ones, have access to the smaller and older parks in preferred sectors. Most small and medium enterprises are dispersed across cities in small organic clusters that the government has not nurtured.

In Sub-Saharan Africa, distorted (or missing) land markets and an inefficient construction industry make developing industrial land costly. So, governments are turning to industrial parks, financed and developed by foreign investors, mostly from China. Most domestic firms outside the parks are located randomly, many in small organic clusters. By offering better industrial land in the zones, Tanzania has attracted some foreign direct investment in agroprocessing.

Land for the Backward Integration of Supply Chains

The comparative value chain analysis identifies land as a key input for the production of heavy-weight, high-volume, farm-grown raw materials for light manufacturing such as agroprocessing (dairy and wheat), leather (ranches where cattle are bred for the meat and leather industries), and wood (planned afforestation; see Global Development Solutions 2011). The cost savings from the large-scale production of local raw materials can make a significant difference in the competitiveness of light manufacturing.

For landlocked Ethiopia and Zambia, the gains from co-location of materials production and processing increase with distance from the coast. But for coastal countries, too, the potential gains from the backward integration of the supply chain for local raw materials are large. Sourcing domestic raw materials requires commercial land, but the absence of a land market in Ethiopia hinders commercial farming and forestry. Land policy limits the entry of large farmers: Ethiopia has only two or three large farms with industrial operations. The success associated with exceptional cases—notably rose plantations that now employ as many as 50,000 workers—illustrates the large potential payoff to enterprise-friendly reform of landholding arrangements.

Land for Affordable Housing

Shortages of residential housing and affordable transport for industrial workers in Sub-Saharan Africa lead to some combination of higher wages (to attract workers) and higher living costs (to travel to and from distant workplaces). To stay close to their workers (and customers) who cannot afford long commutes, small formal metal producers in Dar es Salaam and Addis Ababa prefer to risk fines for operating workshops encroaching on the sidewalks or in crammed compounds rather than relocate to larger factories in peripheral areas, which lack affordable residential space and good transport. And with traffic congestion, commuting shortens working hours.

Worker housing is also a problem in Vietnam, but public transport allows large numbers of workers to commute over fairly long distances to reach even the larger industrial parks on the periphery of Hanoi and Ho Chi Minh City. China's state enterprises have a long history of building housing in or adjacent to factory compounds. Despite large-scale privatization of urban housing beginning in the 1990s, local governments often build housing adjacent to factory sites to accommodate migrant workers, even in less prosperous inland provinces such as Jiangxi. Equipped with the basic amenities, hostels are often free for employees. The proximity to work saves transport costs. This policy helps to contain wage pressures and allows firms to compete on low profit margins.[4] It also allows workers to save a very high proportion of their wages, which they often use later as capital to open new businesses. Large-scale migration of workers, which has moved millions of Chinese far from their original homes, often involves lengthy family separations. In China, workers willingly shoulder the burden of these social costs in pursuit of higher incomes.

Marketing of Goods: Land for Warehousing, Showrooms, and Trading Inputs and Outputs

The qualitative interviews show that few small and medium enterprises in Sub-Saharan Africa have access to land for business transactions. Without storage space, owners purchase inputs at retail rather than wholesale prices. Even in one of Addis Ababa's wood clusters on government-allocated land, the 35 smaller firms are unable to find an individual or common storage facility. The same constraint for grain-milling firms leads to high spoilage rates due to mice and rain. The paucity of showroom space forces small producers to manufacture products only on order. Showrooms would facilitate information sharing between customers and producers on variety, quality, fashions, and designs. But in Sub-Saharan Africa's small and medium enterprises, customers can choose only from photographs of products made for someone else. To obtain new orders, producers rely mostly on repeat customers and word-of-mouth references.

Sub-Saharan Africa's producers do not have easy access to a variety of high-quality inputs and cannot readily purchase large volumes of raw material to satisfy large and time-sensitive orders. The absence of input markets raises the cost of information and search and undermines customer-seller relationships that evolve through frequent interactions and facilitate the delivery of large orders on time.

These are exactly the sort of difficulties that development-oriented local Chinese governments have addressed with great vigor. Previously cited examples in Yiwu, Zhejiang Province (see Ding 2010) and Xinji, Hebei Province (see Blecher and Shue 2001) could be replicated many times over. They show that local governments identified opportunities for profitable expansion, pushed local businesses to adopt measures needed to grasp such opportunities, constructed suitable facilities, and provided (or arranged to obtain) suitable information and advice.

A Possible Solution from Asia

This section considers possible solutions from China's plug-and-play industrial parks and Africa's experience so far with these parks.

China's Plug-and-Play Industrial Parks Catering to Small and Large Enterprises

Many studies have documented the contributions of China's coastal special economic zones as platforms for attracting export-driven foreign direct investment and as testing grounds for key reforms. The Shenzhen Special Economic Zone next to Hong Kong SAR, China, transformed a fishing village into a leading global city of 8 million people in less than 30 years. China's smaller industrial parks are less well known, but they too have contributed substantially to the nation's astonishing industrial development.

China has more than 1,000 industrial parks driven by a central government policy encouraging their development. Many cities and counties have emulated the large zones of the central and provincial governments. Local governments develop industrial parks to spur local growth and increase tax revenues (Li and Zhou 2005). The parks have enabled many small and medium Chinese enterprises to grow from family operations focused on domestic markets into global powerhouses. But not all Chinese industrial parks have been successful; the better ones were built on existing or potential industrial strengths—in other words, on local comparative advantage.

China's successful industrial parks provide enterprises with security, good basic infrastructure (roads, energy, water, sewers), streamlined government regulations (through government service centers), and affordable industrial land. They also provide technical training, low-cost standardized factory shells, and free and decent housing for workers next to the plants. By helping small Chinese enterprises to grow into medium and large enterprises, the country has avoided a shortage of medium firms—the "missing middle"—evident in most Sub-Saharan countries. China's parks focus on specific industries, such as leather and textiles in Nanchang, furniture in Ji'an, and electronics in Ganzhou (for further examples, see Zeng 2008; Sonobe and Otsuka 2006; Sonobe, Hu, and Otsuka 2002).

More advanced industrial parks offer market analysis, accounting, import and export information, and management advice and help firms to recruit and train workers. For example, parks near the Yangtze River delta place a strong emphasis on helping firms to get business licenses and hire workers. Parks may also have facilities to address environmental challenges.

The plug-and-play industrial parks have greatly reduced start-up costs and risks for small and medium enterprises that have sufficient scale, capital, and growth prospects to take advantage of larger facilities at a phase in their development when they are unable to obtain bank loans. They have also facilitated industrial clusters, generating substantial spillovers as well as economies of scale and scope for Chinese industries. The clusters are facilitated further by government support for input and output markets.

In a nutshell, Chinese governments, especially at the local level and particularly in the central and southeastern coastal provinces, have taken the initiative and moved energetically to foster the development of small and medium enterprises by providing public goods and market information about suppliers and buyers rather than large-scale subsidies. Gradual reforms have pushed the cost of energy and utilities in the direction of market prices; firms that fail to pay their bills cannot expect to have continued access to electricity and other utilities. Competition is intense among firms in light manufacturing (and in many other sectors of China's economy). China's economic reforms have vastly reduced opportunities for failing firms, especially small and medium operations, to receive government bailouts. China now has several hundred million migrant workers, with the largest flow of migrants from west to east. China's coastal cities have become magnets for millions of migrant workers, many of whom live and work in various types of special zones and industrial parks. The design and operation of these zones and parks facilitates employers' access to large numbers of workers and contributes to the migrants' efforts to achieve their own goals of upward mobility through hard work and diligent saving. Many Chinese migrants remit substantial funds to their original homes to support parents, spouses, and children left behind, to build new housing, and sometimes to support future business plans.

In Africa, by contrast, workers spend the bulk of their incomes on housing, food, and transport and achieve much lower savings rates than their Chinese counterparts. Reforming legal and administrative provisions that limit access to land, especially industrial plots connected to utilities and convenient transportation, will encourage the growth of manufacturing enterprises that can raise the productivity, and hence the wages, of large numbers of workers. In this way, growing numbers of low-income Africans can begin to access the sort of benefits that have improved the lives and prospects of many millions of Chinese workers during the past several decades of reform.

Policy Recommendations for Africa

African governments should develop such plug-and-play industrial parks next to main cities and ports. This would eliminate in one stroke the very high inland transport costs for exporters in Africa.

As in China, industrial parks can bypass difficult land reform, which can take years. African governments can test a variety of policies before gradually applying them more widely. The demonstration effects can overcome political economy constraints.

But industrial zones are not new to Africa. Why have they failed (box 3.1), and why would they work now?

BOX 3.1

Why Have Industrial Parks Failed in Africa?

Industrial parks have played a catalytic role in facilitating industrial upgrading and export-led growth in East Asia, most notably in "tiger economies" during the 1980s and in China since the early 1990s, but also in Latin America and parts of South Asia. The African experience with industrial parks over the past two decades, which has mostly involved traditional export-processing zones, has been less spectacular. With the significant exception of Mauritius and the partial (initial) success of Kenya, Madagascar, and Lesotho, most African zones have failed to attract significant investment, promote exports, or create sustainable employment. Only Mauritius has used industrial parks as an effective vehicle to support economic transformation.

While many factors that contributed to the failure of individual zones are very specific, some have held back the potential of industrial parks across Africa:

- *Poor strategic planning: a mismatch with comparative advantage.* One of the striking features of many zones in Africa is that the few investors they have managed to attract are spread across a wide range of manufacturing sectors. This contrasts with hugely successful Asian examples like the Hsinchu Science and Industrial Park in Taiwan, China (see Hsueh, Hsu, and Perkins 2001; Amsden and Chu 2003) that set out to build industry clusters. One of the reasons is that many zones have been initiated without careful studies of market demand or strategic planning. They are poorly targeted (often at sectors well outside the country's comparative advantage), and few have been integrated into their country's broader economic policy framework.

- *Poor choice of location.* Zone location is determined too often by political rather than commercial or economic considerations. Rather than co-locating zones with key gateway infrastructure, many countries have attempted to locate at least one zone in a "lagging" or remote region. And few have done enough to address the infrastructure connectivity, labor skills, and supplies that these regions lack.

- *Insufficient investment in infrastructure.* Some African zones offer infrastructure inside the zones that, while usually not world class, is of higher quality than is typically available in the domestic economy. But infrastructure in many zones mirrors the worst domestic experiences, including water shortages, electricity outages, and health, safety, and environmental shortfalls. Zones that lack even basic internal infrastructure have little chance of success. Moreover, in many cases infrastructure stops at the zone gates, with investors having to deal with poor roads, port-related delays, and little access to social infrastructure.

- *Poor implementation capacity and lack of authority.* The authorities responsible for developing, promoting, and regulating African special economic zones often lack resources as well as the institutional authority and implementation capacity to carry out their mandates. The resulting confusion in procedures and controls—most notably in customs administration—undermines the competitiveness of many African industrial parks.

- *Lack of high-level support and policy stability.* Many African countries have shown only a half-hearted commitment to zones, passing zone laws but not implementing regulations or providing adequate resources for management, infrastructure, and promotion. Fundamental "rules of the game" often change from year to year, and many programs have suffered from poor coordination of trade policy and a failure to establish a policy environment that offers investors transparency and predictability.

In addition, African industrial parks may be lagging because of poor timing. The rapid growth of industrial parks on a global basis and their success in contributing to export-led growth benefited from an unprecedented era of globalization in the 1980s and 1990s and the rise of global production networks. But African countries, most of which only launched programs well into the 1990s and 2000s, face a much more competitive environment, with the emergence and entrenchment of "factory Asia," the expiration of the Multifiber Arrangement, the consolidation of production networks, and the recent slowdown in demand in traditional export markets. Some countries may have developed the wrong zone model at the wrong time.

Source: Farole 2011.

Note: Lesotho does not have a zone program, but it does combine the same policy instruments to support export manufacturers, including a special fiscal and administrative regime along with public provision of industrial infrastructure.

Notes

1. While China did experience some land issues, particularly at the local levels regarding the compensation paid to farmers for their land, these issues were relatively small compared to the progress made in making industrial land available to entrepreneurs.
2. For a review of policies, investment incentives, and performance of policy in support of the development of the manufacturing industry in Vietnam, see Le (2010).
3. Around 1998 the government designed special fiscal incentives for new and existing firms to locate in remote land development projects for ethnic minority areas, mountain areas, islands, and other underdeveloped areas. A preferential system was set up to assist domestic investors by allotting or renting land according to land legislation and building the industrial zone's infrastructure for renting them. After 1998 domestic investors also benefited from reductions in the land use tax.
4. In both Vietnam and China most firm owners also provide free meals. While this is an innovative solution to one of the leading constraints on firm expansion, hostels that separate the worker from the family for long periods have high social costs. Unless this model is adapted to the local socioeconomic conditions, policy makers and workers in countries with different political systems and beliefs may have reservations about it.

References

Amsden, Alice H., and Wan-wen Chu. 2003. *Beyond Late Development: Taiwan's Upgrading Policies.* Cambridge, MA: MIT Press.

Blecher, Marc, and Vivienne Shue. 2001. "Into Leather: State-Led Development and the Private Sector in Xinji." *China Quarterly* 166 (2001): 368–93.

Ciccone, Antonio, and Robert E. Hall. 1993. "Productivity and the Density of Economic Activity." *American Economic Review* 86 (1): 54–70.

Ding, Ke. 2010. "The Role of the Specialized Markets in Upgrading Industrial Clusters in China." In *From Agglomeration to Innovation: Upgrading Industrial Clusters in Emerging Economies,* ed. Akifumi Kuchiki and Masatsugu Tsuji, 270–89. Houndmills, U.K.: Palgrave Macmillan.

Dollar, David, Mary Hallaward-Driemeier, and Taye Mengistae. 2004. "Investment Climate and International Integration." Working Paper 3323, World Bank, Washington, DC.

Fan, C. Cindy, and Allen Scott, 2003. "Industrial Agglomeration and Development: A Survey of Spatial Economic Issues in East Asia and a Statistical Analysis of Chinese Regions." *Economic Geography* 79 (3): 295–319.

Farole, Thomas. 2011. "Special Economic Zones in Africa." Poverty Reduction and Economic Management, International Trade, World Bank, Washington, DC.

Global Development Solutions. 2011. "The Value Chain and Feasibility Analysis; Domestic Resource Cost Analysis." Background paper (Light Manufacturing in Africa Study). Available online as Volume II at http://worldbank.org/africamanufacturing. World Bank, Washington, DC.

Hausman, Warren H., Hau L. Lee, and Uma Subramanian. 2005. "Global Logistics Indicators, Supply Chain Metrics, and Bilateral Trade Patterns." World Bank Policy Research Working Paper 3773, Washington, DC.

He, Canfei, Yehua Dennis Wei, and Xiuzhen Xie. 2008. "Globalization, Institutional Change, and Industrial Location: Economic Transition and Industrial Concentration in China." *Regional Studies* 42 (7): 923–45.

He, Canfei, and Junsong Wang. 2010. "Spatial Restructuring of Chinese Manufacturing Industries and the Productivity Effects of Industrial Agglomerations." Background Paper for *Innovation Policy Note*, ed. Hamid Alavi. College of Urban and Environmental Science, Peking University; Peking-Lincoln Institute Center for Urban Development and Land Policy.

He, Canfei, and Shengjun Zhu. 2007. "Economic Transition and Regional Industrial Restructuring in China: Structural Convergence or Divergence?" *Post Communist Economics* 19 (3): 321–46.

Hsueh, Li-min, Chen-kuo Hsu, and Dwight H. Perkins, eds. 2001. *Industrialization and the State: The Changing Role of the Taiwan Government in the Economy, 1945–1998.* Cambridge, MA: Harvard University Press.

Lall, Sanjaya. 2004. *Reinventing Industrial Strategy: The Role of Government Policy in Building Industrial Competitiveness.* G-24 Discussion Paper Series 28. United Nations, New York.

Le, Duy Binh. 2010. "Review of Policies and Investment Incentives: Performance of Policy in Support of the Development of the Manufacturing Industry in Vietnam." World Bank, Hanoi.

Lee, Yung Joon, and Hyoungsoo Zang. 1998. "Urbanization and Regional Productivity in Korean Manufacturing." *Urban Studies* 35 (11): 2085–99.

Li, Hongbin, and Li-An Zhou. 2005. "Political Turnover and Economic Performance: The Incentive Role of Personnel Control in China." *Journal of Public Economics* 89 (9/10): 1743–62.

Limão, Nuno, and Anthony J. Venables. 2001. "Infrastructure, Geographical Disadvantage, Transport Costs, and Trade." *World Bank Economic Review* 15 (3): 451–79.

Pan, Zuohong, and Fan Zhang. 2002. "Urban Productivity in China." *Urban Studies* 39 (12): 2267–81.

Sonobe, Tetsushi, D. Hu, and Keijiro Otsuka. 2002. "Process of Cluster Formation in China: A Case Study of a Garment Town." *Journal of Development Studies* 39 (1): 118–39.

Sonobe, Tetsushi, and Keijiro Otsuka. 2006. "The Division of Labor and the Formation of Industrial Clusters in Taiwan." *Review of Development Economics* 10 (1): 71–86.

Wang, Jici. 2001. *Innovative Space: Industrial Cluster and Regional Development.* Beijing: Peking University Press.

Zeng, Douglas. 2008. "Innovation and Cluster Development in China." World Bank, Washington, DC. http://www.clusteringconference.com/html/TR/sunumlar/Session%20V/P2-Dougles%20Z.%20Zeng%20(Session5).pdf.

Chapter 4

Finance

The main source of financing is retained earnings in all five study countries, and the need for capital investments is relatively small in light manufacturing. But access to finance is an important constraint across all light manufacturing sectors, especially for small and medium enterprises. All firms require additional resources to invest in technology, improve buildings, and buy new land. The high cost of formal finance is driven in part by difficulties in using assets as collateral, especially for firms in agribusiness, because formal banks will not accept agricultural produce or livestock as collateral.

Access to Finance and Firm Performance

Financial access variables have a significant effect on firm growth. With other factors held constant, having a loan or overdraft facility increases the growth in a firm's number of permanent employees by 3.1 percent, being credit constrained reduces a firm's employment growth by 1.9 percent, having sales credit increases a firm's growth by 2.6 percent, and having external investment funds increases growth by 4.2 percent.[1] These strong results show that access to finance indeed matters for firm growth.

But there is a difference in the start-up and operating phases of a firm. Start-up finance for firms in light manufacturing is, in the vast majority of cases, from own savings, friends, and family. In China many small and medium enterprises revealed that they used savings from migrant work to start their operations. Little evidence was found that formal sources of finance are significant at the start-up phase. Formal bank finance becomes important when a micro or small firm wants to expand, upgrade technology, or increase production. In Sub-Saharan Africa firms have to find significant up-front capital to buy land, build factory premises, and invest in machinery. The cost of finance and the requirement for collateral prevent them from getting loans to finance expansion.

When we consider the transformation of successful small firms into medium or large firms, lack of formal financing options is a key constraint in Sub-Saharan Africa. Difficulties in accessing finance can contribute to the "missing middle" phenomenon, leaving small enterprises trapped in low technology and low pro-

Table 4.1 Percentage of Firms That Could Borrow to Purchase Additional Machinery, Equipment, or Vehicles in the Five Countries, by Source

Source	China	Vietnam	Ethiopia	Tanzania	Zambia
Bank	36	34	8	60	32
Nonbank financial institution	10	1	45	12	20
Government agency	0	1	7	3	12
Family or friends	11	22	4	5	9
Moneylender	1	4	0	4	1
Other	0	0	0	0	7
Number of observations	303	300	250	262	263

Source: Fafchamps and Quinn 2011.

ductivity, without the means to upgrade skills and technology. Low financial sector development affects firm size and skews the distribution toward small and medium firms, especially among firms that perceive access to finance as an obstacle (Dinh, Mavridis, and Nguyen 2010).

Our quantitative survey asked firms whether they *could* borrow from various sources to purchase additional machinery, equipment, or vehicles. In China and Vietnam, around 35 percent of firms said they could borrow from a bank, but only 8 percent of small and medium enterprises reported this for Ethiopia. In contrast, 60 percent of small and medium enterprises in Tanzania and 32 percent of those in Zambia reported that they could borrow from a bank (table 4.1). But this is contrary to data showing that only 4 percent of firms in Tanzania and 3 percent in Zambia actually borrowed from a bank. Why have small and medium enterprises in Sub-Saharan Africa not used bank finance, even when they perceive that it is available to them?

The Availability and Cost of Finance

More than 80 percent of small and medium enterprises surveyed used retained earnings to finance their last purchase of machinery and equipment (table 4.2). This confirms the recurring response in qualitative interviews that own savings for the most part financed firm expansion. But as firms grow, they use more diverse sources, especially in China.

For innovations in new products, production processes, or delivery systems, the quantitative survey indicates that in all countries except China, small and medium enterprises require additional finance. After retained earnings, the main source of financing for innovation is friends and relatives (table 4.3), cited twice as often by Asian as by African respondents. Banks are cited by a minority of respondents, but more frequently by Asian than by African firms (19 percent for China, 1–11 percent for Sub-Saharan Africa). New capital from existing

Table 4.2 Source of Funding for the Purchase of Machinery, Equipment, or Vehicles in the Five Countries
% of firms

Source	China	Vietnam	Ethiopia	Tanzania	Zambia
Internal funds or retained earnings	80	82	88	80	86
Bank	23	18	2	4	3
Nonbank financial institution	7	2	10	2	1
Government agency	0	0	0	0	0
Family or friends	7	18	4	3	1
Hire-purchase or credit from the equipment supplier	5	1	0	0	0
Other	0	1	3	1	8
Number of observations	265	300	250	262	250

Source: Fafchamps and Quinn 2011.

Table 4.3 Source of Funding for Innovation in the Five Countries
% of firms

Source	China	Vietnam	Ethiopia	Tanzania	Zambia
Retained earnings	55	87	88	47	79
Friends and relatives	16	16	6	8	8
Bank	19	16	4	11	1
New capital from owners	12	9	9	2	5
Customers	14	4	9	6	3
Financial institution	23	1	7	0	2
Raw material supplier	4	8	2	0	0
New owners	4	1	0	0	0
Domestic joint venture	3	0	0	0	0
Equipment supplier	3	0	0	0	0
Government agency	1	0	0	0	0
Foreign joint venture	0	1	0	0	0
Development agency	0	0	0	0	0
Other	0	0	2	0	5
Number of observations	73	121	113	66	99

Source: Fafchamps and Quinn 2011.

owners is also cited on average more often by Asian respondents. In addition, Chinese respondents mention advances from customers and credit from financial institutions as significant sources, largely omitted by other respondents.

Generally, participation in the formal financial system is higher among firms in China. To access formal bank finance, the firm needs to establish a relationship before lending through a bank account. Smaller Chinese firms are substantially more likely to have a bank account than firms in other countries (table 4.4).[2]

Table 4.4 Percentage of Firms That Have Borrowed in the Five Countries, by Source, 2006–10

Source	China	Vietnam	Ethiopia	Tanzania	Zambia
Bank	33	36	4	3	3
Nonbank financial institutions	15	4	32	2	4
Government agency	12	2	0	0	1
Family or friends	15	19	18	2	3
Moneylender	0	5	0	0	1
Other	0	0	6	0	1
Number of observations	303	300	250	262	263

Source: Fafchamps and Quinn 2011.

Firms in China and Vietnam borrow mostly from banks, but Chinese firms also obtain formal financing through nonbank financial institutions and government agencies. Very few firms in Tanzania and Zambia borrow, indicating that formal lending sources fail to meet the needs of small and medium enterprises in those countries.

Among firms in Sub-Saharan Africa that do borrow, the costs and collateral requirements are significantly higher than for their counterparts in Asia. Small and medium enterprises in our quantitative survey pay an average annual interest rate of about 4.7 percent in China, compared with about 10 percent in Ethiopia, 14 percent in Vietnam and Tanzania, and 21 percent in Zambia. Firms reported that 173 percent of the loan amount is required as collateral in Ethiopia (2006), 146 percent in Zambia (2007), and 124 percent in Tanzania (2006). This proportion was 88 percent in China (2003).

Working capital allows small and medium enterprises to take on bigger orders, be more responsive to customers, and remain liquid. Among firms with an account, few have an overdraft facility outside China. In Vietnam and Ethiopia only a handful of manufacturing firms have an overdraft facility, rising to 6 percent in Tanzania, 19 percent in Zambia, and 63 percent in China.

Overdraft facilities appear to help Chinese manufacturing firms to obtain short-term finance at a median annual interest rate of 7.5 percent. Only about 20 percent of the firms said they are required to provide collateral (the median collateral is 65 percent of the value of the overdraft facility).[3]

Why Is the Cost of Formal Finance So High and the Availability So Constricted?

There are a number of reasons for the high cost and limited availability of finance in Sub-Saharan Africa, including issues of risk, weak asset markets, and low savings rates, which are discussed in detail below.

High Risk

The risks associated with operating a light manufacturing business in Sub-Saharan Africa are very high, as shown by the constraints discussed in this book. For these very reasons, banks will not lend to firms and sectors in which risks are too high and profits are too low. Moreover, the low savings rate and distortions in the credit market increase the costs of finance and restrict access to formal finance. The costs are elevated due to information asymmetries (lack of credit registries), market uncertainty (dependence on imported inputs, macroeconomic and political environment, weak infrastructure), and the risk of firm failure (weak entrepreneurial skills, poor market access, access to new technology). The competitive weakness of most small and informal firms also makes it difficult for banks to assess creditworthiness.

Weak Asset Markets

A major barrier for young firms seeking to access finance is the higher collateral requirement for loans in Africa than in China. Although small and medium firms in China rarely receive direct financing from government, official efforts to provide factory shells and access to land in industrial zones allow successful firms to expand without bearing the time and expense needed to build their own factory; such arrangements allow growing firms to conserve funds that may then be used as collateral to obtain loans. In addition, recent policy initiatives may have begun to erode long-standing barriers that effectively deny bank credit to most small and medium enterprises.

Land policy in Ethiopia, Tanzania, and Zambia makes it is difficult for banks to accept land as collateral. In Ethiopia the government has stipulated that land can be sold only if it has been "developed," meaning that the owner must build a structure on the land. In addition, banks do not consider the viability and track record of a business in assessing a loan application. Although the leasing of machinery is allowed by law, banks do not accept machinery as collateral.

Even controlling for firm size, Chinese firms face substantially lower average collateral requirements than their counterparts in Tanzania and Zambia (figure 4.1). As shown in the interviews, the cost of finance is rarely the issue. The problem is the difficulty of using land or assets as collateral. Small and medium enterprises in China are better able to obtain formal finance than firms in Africa because the requirement for collateral is lower.

This advantage is evident in agroprocessing. The barriers to small Ethiopian dairy farmers using land or livestock as collateral reduce their ability to upgrade production technology or compete even domestically. The prohibitive cost and low access to finance prevent farmers from upgrading their cattle to high-yielding cross-breeds, and the use of low-yield cattle further undermines their ability to compete.

Figure 4.1 Collateral Requirements in the Five Countries

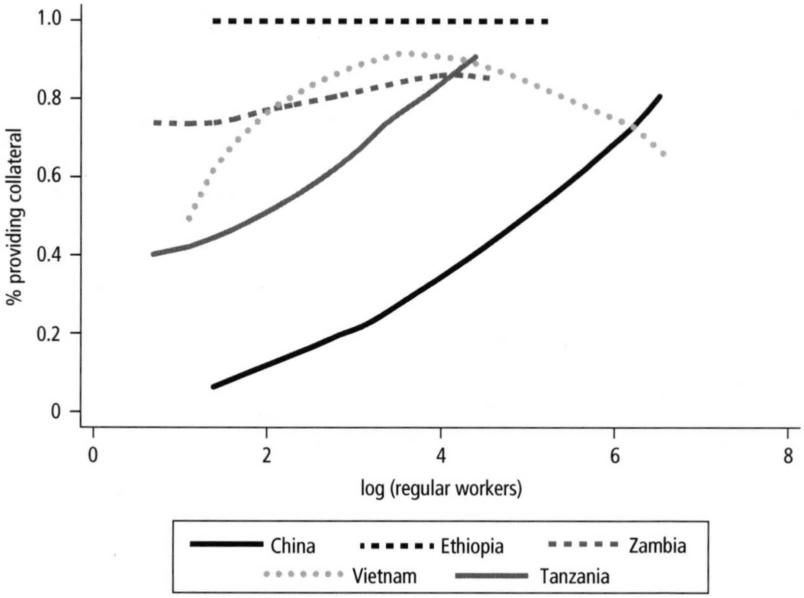

Source: Fafchamps and Quinn 2011.

Low Savings Rates

Gross domestic savings have been consistently higher in China (50 percent of gross domestic product [GDP]) than in Vietnam (30 percent) and Sub-Saharan comparators (20 percent; figure 4.2). Low domestic savings limit the availability of capital for private investment. Some would argue that a low savings rate is a binding constraint on private sector growth. Although the correlation between these two factors is strong, the direction of causality is debatable. As economies grow, individual and national incomes rise, the proportion of income needed for immediate consumption falls, and domestic savings rates are likely to increase.

Possible Solutions from Asia

Banks rarely lend to a firm without some form of guarantee or collateral. Because of the difficulties in buying and selling land, banks in Sub-Saharan Africa do not accept land as collateral. As mentioned, the government of Ethiopia has stipulated that land can be sold only if it has been "developed." But

Figure 4.2 Gross Domestic Savings as a Percentage of GDP in the Five Countries, 1985–2009

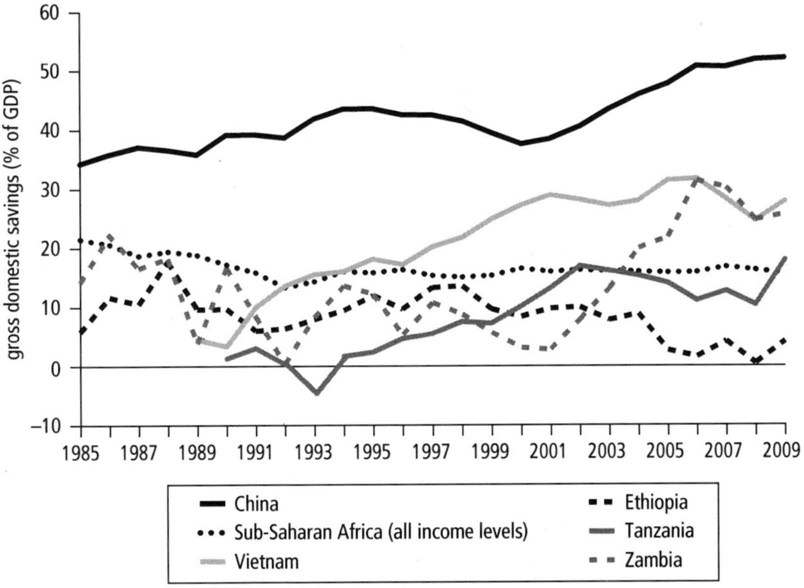

Sources: World Development Indicators and Global Development Finance databases, various years.

there is uncertainty over what "developed" means in practice, and banks often refuse to accept agricultural land as collateral. Since banks also decline to allow farmers to offer cattle as collateral, Ethiopian bank policy effectively restricts the expansion of dairy farming and milk processing. Land reforms have the potential to free up the access of private firms to long-term leases for industrial and agricultural land; they could also encourage the formalization of rental arrangements to facilitate the use of land as collateral against bank loans.

Support to nonbank financial institutions could increase the use of leasing, factoring, and other products by enterprises, especially small and medium enterprises, in the absence of a developed financial system (see Beck and Demirgüç-Kunt 2006). Leasing is a useful tool for upgrading equipment without collateral or guarantees. A legal and institutional framework needs to be in place to support the establishment of leasing and factoring companies.

The limited second-hand market for machinery in many Sub-Saharan African economies reduces the viability of banks using machinery as collateral. The government could step in to provide modest incentives for banks and other financial institutions to offer financing for machinery to well-managed firms.

As the industry grows, the market for second-hand machinery will expand, making its use as collateral more efficient.

Recent analysis of the role of clustering in manufacturing development in China highlights the positive effects of clustering for access to finance (Long and Zhang 2011). First, this production model involves many small firms and a high division of labor, rather than large firms that integrate all stages of production. This structure reduces the capital investment required for each firm to operate. Individual firms still face financial constraints, but the required capital is lower. Second, in the early stages of China's development, small firms did not use formal finance. The cluster structure created social capital, which supported trade credit and informal finance.

The positive spillovers from improving access to finance for small and medium enterprises in light manufacturing provide more justification for government to support both industrial clusters for small firms and industrial parks for larger firms (see chapter 3).

Notes

1. If we include all of these significant financial access variables in one model after controlling for firm characteristics, they still have significant effects on employment growth, although the effects are smaller. And if we use the same model and run the regressions in different regions, the significance and signs of the effects remain the same (Dinh, Mavridis, and Nguyen 2010).
2. Asian firms are, on average, 10 percentage points more likely to have an account.
3. But nonresponse may be an issue. Of the 160 Chinese firms reporting an overdraft, only 119 answered the question about collateral and only 68 answered the question about the interest rate.

References

Beck, Thorsten, and Aslı Demirgüç-Kunt. 2006. "Small and Medium-Size Enterprises: Access to Finance as a Growth Constraint." *Journal of Banking and Finance* 30 (11): 2931–43.

Dinh, Hinh T., Dimitris Mavridis, and Hoa B. Nguyen. 2010. "The Binding Constraint on Firms' Growth in Developing Countries." Background paper (Light Manufacturing in Africa Study). Available online in Volume III at http://econ.worldbank.org/africa manufacturing. World Bank, Washington, DC.

Fafchamps, Marcel, and Simon Quinn. 2011. "Results from the Quantitative Firm Survey." Background paper (Light Manufacturing in Africa Study). Available online in Volume III at http://econ.worldbank.org/africamanufacturing. World Bank, Washington, DC.

Long, Cheryl, and Xiaobo Zhang. 2011. "Cluster-Based Industrialization in China: Financing and Performance." *Journal of International Economics* 84 (1): 112–23.

Chapter 5

Trade Logistics

Poor trade logistics penalize firms that rely on imported inputs and doubly hit exporters (mostly large and medium firms in Africa). On average, they add roughly a 10 percent production cost penalty across the five subsectors in the three African countries (figure 5.1).

Poor trade logistics also cause long and uncertain delays, which are not acceptable to most global buyers, especially in the time-sensitive apparel industry. As a result, production in Ethiopia and Tanzania is confined mostly to small market niches. Ethiopia exports small volumes of low-value products—the free-on-board (f.o.b.) price for an Ethiopian polo shirt is around US$3.20, much lower than the US$5.50 price of an equivalent Chinese polo shirt. The higher Chinese f.o.b. price results from higher-quality shirts and the premium that global buyers put on China's capacity to offer greater choice, bigger volumes, and shorter and more certain delivery times. Tanzania exports polo shirts at an f.o.b. price similar to China's, but these are small-volume specialty products with orders as low as 1,000 pieces per style, not the fairly standard orders of 15,000–60,000 pieces. The small orders mean higher input costs, lower capacity use, and higher overheads.

Why Are Trade Logistics So Important to Competitiveness in Light Manufacturing?

Good trade logistics connect low-income countries to global output and input markets. Both China and Vietnam have built their export sectors on the back of very good trade logistics; establishing industrial export zones next to efficient ports greatly aided their efforts to expand the sale of manufactures to overseas markets.

But in Africa poor trade logistics increase production costs (in most cases wiping out the labor cost advantage) and lead to long and unreliable delivery times, which make it extremely difficult for firms to fill just-in-time orders. Therefore, global buyers are hesitant to place large orders.

Figure 5.1 Impact of Poor Trade Logistics on Total Production Costs, by Sector

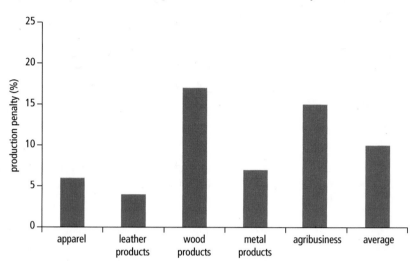

Source: Global Development Solutions 2011.
Note: Apparel and leather products do not include the penalty associated with long and uncertain delivery times (up to 10 percent of production cost in apparel).

Although poor trade logistics can act as natural protection for domestic producers selling to domestic and regional markets in Africa, that protection is moot for the many products for which most inputs must be imported (see chapter 2). Indeed, the cost of importing low volumes of many different types of inputs may exceed the cost of importing the final good. Studies have shown that most African countries are below the global "efficient frontier" for apparel production, as defined by the cost of logistics and labor, although Madagascar and Ethiopia are not too far from it (figure 5.2).

Trade Logistics Performance

According to the World Bank's *Doing Business 2011* (World Bank 2011a), China and Vietnam perform especially well on the trading-across-borders indicators, which show the cost and time of moving a 20-foot container from the port of arrival to the largest business city in the country (excluding ocean travel). China ranks 50 among 183 economies, and Vietnam ranks 63 (table 5.1). The cost to export and import per container ranges from US$500 in China to US$645 in Vietnam. But it can be as high as US$1,200 for exports in Tanzania and US$3,000 for imports in Ethiopia or Zambia. The time to import and export is

Figure 5.2 Logistics Performance Index and Annual Labor Costs per Worker in Apparel Production in Select Countries, 2010

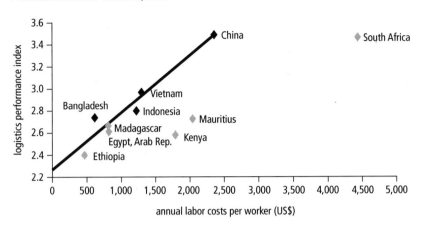

Source: Conway and Shah 2010.
Note: Trend line indicates the average, for example, for a given level of logistics performance; countries beyond the trend line have above-average labor cost.

Table 5.1 Time and Cost of Trading across Borders in the Five Countries

Country	Cost to import a 20-foot container (US$)	Time to import (days)	Cost to export a 20-foot container (US$)	Time to export (days)	Ranking[a]
Ethiopia	2,993	45	1,890	44	157
Tanzania	1,475	31	1,262	24	109
Zambia	3,315	56	2,664	44	150
Sub-Saharan Africa average	2,492	38	1,962	32	
Vietnam	645	21	555	22	63
China	545	24	500	21	50

Source: World Bank 2011b.
a. Ranking among 183 economies on all Doing Business trading-across-borders indicators.

at least twice as long for Ethiopia and Zambia as it is for China and Vietnam. In addition, these data do not reflect the time for customs procedures at Djibouti and Dar es Salaam. Although Tanzania benefits from having its own port, it still ranks far behind China and Vietnam on trading across borders.

On the World Bank's logistics performance index, Tanzania (95), Ethiopia (123), and Zambia (138) sit firmly in the bottom half of the ranking (155 countries ranked in the 2011 index) with China at 27 and Vietnam at 53.

Factors Leading to Poor Trade Logistics in Africa

Poor trade logistics in Africa result from four broad factors (table 5.2).

Higher Inland Transport Costs

For Ethiopia and Zambia higher inland transport costs add more than a 2 percent production cost penalty in the apparel sector and a 10-day delay, due to longer distances, inadequate transport infrastructure, and a lack of competition in the trucking industry (Teravaninthorn and Raballand 2008). Addis Ababa is more than 800 kilometers from the port of Djibouti. Zambia's shortest route to the sea is through Zimbabwe to the Mozambican port of Beira. Most Zambian import and export traffic is now handled by the port of Durban, more than 2,300 kilometers from Lusaka by road. The rehabilitation of the railway between Lusaka and Beira and of the Beira port will be essential to alleviating Zambia's trade logistics disadvantage. Longer distances and inadequate transport infrastructure add a 2 percent production cost penalty for apparel in Ethiopia and Zambia.

Higher Port and Terminal Handling Fees

Although Ethiopia now enjoys access to the new state-of-the-art Doraleh container port at Djibouti, port and terminal handling fees remain high. For Tanzania, insufficient storage capacity, poor container management, and a distorted profit incentive of the container terminal operator (storage charges are a main source of revenue) have led to severe congestion at Dar es Salaam. Some shipping companies have reduced the frequency at which their ships stop in Dar es Salaam, with one dropping its route altogether. Shorter turnover times (with custom reforms) together with the expansion of the Tanga port should alleviate congestion. Higher port and terminal handling fees in Africa add a 1 percent production cost penalty for apparel.

Table 5.2 Four Main Factors That Impair Trade Logistics in the Five Countries
US$ per 20-foot container

Country	Inland transportation	Port and terminal handling	Customs clearance	Preparation of documents and letters of credit	Total
Ethiopia	1,000	500	340	600	2,440
Tanzania	200	400	250	520	1,370
Zambia	2,300	300	110	280	2,990
Vietnam	230	160	100	110	600
China	120	80	70	250	520

Source: World Bank 2011b.

Higher Customs Clearance and Technical Control Fees

Lack of automation and a single customs window result in higher custom fees and delays, which further compound the African deficit in infrastructure by reducing the turnover of containers and ships. This adds a production cost penalty of a little less than 1 percent for apparel.

Higher Costs of Document Preparation and Letters of Credit

Another 2 percent production cost penalty (apparel sector) is incurred because of the higher cost of document preparation (including letters of credit). In Ethiopia commercial banks charge 3 percent of the value of the shipment on imports and a 2 percent advisory fee on exports (compared with less than 1 percent in China). The Doing Business indicator probably underestimates the severity of the issue because it is based on a very low container value (US$20,000); the value of a 20-foot container filled with 20,000 polo shirts is US$100,000. The World Bank is planning a study to understand why letters of credit are so expensive in Africa.

Other Factors

Other factors, not captured by the Doing Business indicators, add even more to trade logistics costs in Africa:

- *High cost of foreign exchange.* The National Bank of Ethiopia charges a 1.5 percent foreign exchange commission fee on the dollars needed to import the inputs (a 1 percent production cost penalty in apparel). Waiting for the National Bank's authorization can take up to six months when foreign exchange is scarce.

- *High shipping costs to and from Africa.* It costs 60 percent more to ship to the United States from Djibouti than from China and about the same to ship to Europe, despite the much greater distance from China. One reason is less traffic in Djibouti; another is the monopoly of the state-owned Ethiopian shipping line. Higher shipping costs add a 0.5 percent production cost penalty for apparel.

Improving trade logistics is thus a priority for African countries seeking to compete in export-oriented light manufacturing. Not only does the 10 percent production cost penalty due to poor trade logistics wipe out Africa's labor cost advantage in most sectors, but long and uncertain delays keep African countries out of the higher-value time-sensitive segments of light manufacturing. China and Vietnam started their light manufacturing journey by establishing good-practice industrial zones next to efficient ports. Low-wage African countries with easy access to a port should do the same, as this would also benefit low-skilled workers in landlocked African countries who would be able to migrate to these industrial zones.

Possible Solutions from Asia

China's industrial growth benefited from a large domestic market for firms to exploit economies of scale before entering global trade. Vietnam also benefited from the Association of Southeast Asian Nations market. African governments should explore the potential of regional markets to provide the same benefits for domestic manufacturing firms.

First, African governments should harmonize and improve customs operations by simplifying procedures and leveraging information technology (see de Wulf and Sokol 2005). Delays in customs procedures incur storage costs for containers waiting at the border to be cleared. And the delays and uncertainty over time for customs clearance damage the reputation of local firms and reduce the f.o.b. prices that they can negotiate with global buyers.

Second, African countries should develop the hard infrastructure (across boundaries) to support multimodal systems combining trucking, railways, airways, and shipping to improve connectivity and increase competition. This would entail rehabilitating the railway line between Addis Ababa and Djibouti (planned), expanding the capacity of the port in Dar es Salaam, and rehabilitating the trade corridors between Lusaka and Nacala.

Third, African governments need to increase competition among freight forwarders, shipping companies, and trucking companies by removing price controls and restrictions on foreign direct investment and on cabotage (Teravaninthorn and Raballand 2008).

Fourth, African governments should develop strategic partnerships along key trade corridors. For example, the Ethiopian government should develop a strategic partnership with Djibouti to optimize port operations and minimize charges. Port operations are a key source of revenue for Djibouti, and Ethiopia accounts for more than 70 percent of the volume. Djibouti's future development thus depends in part on Ethiopia's future success in light manufacturing, just as the success of Hong Kong SAR, China, relied on China's. So the new state-of-the-art container port at Doraleh (managed by Dubai Ports) can be viewed as a strategic asset. A good first step is improving the legal framework that regulates the services of private port operators; its current inadequacy hampers the efficiency, competitiveness, and quality of service of the Djibouti port. The old fee structure for private port operators is not suited to port costs (both the types of costs and their amounts), to the new kinds of traffic, or to the new techniques and methods of transport, handling, and management.

References
Conway, Patrick, and Manju Shah. 2010. "Incentives, Exports, and International Competitiveness in Sub-Saharan Africa: Lessons from the Apparel Industry." World Bank, Washington, DC.

de Wulf, Luc, and José B. Sokol. 2005. *Customs Modernization Handbook.* World Trade and Development Series. Washington, DC: World Bank.

Global Development Solutions. 2011. "The Value Chain and Feasibility Analysis; Domestic Resource Cost Analysis." Background paper (Light Manufacturing in Africa Study). Available online as Volume II at http://worldbank.org/africamanufacturing. World Bank, Washington, DC.

Teravaninthorn, Supee, and Gael Raballand. 2008. *Transport Prices and Costs in Africa: A Review of the Main International Corridors.* Washington, DC: World Bank.

World Bank. 2011a. *Doing Business 2011.* Washington, DC: World Bank.

———. 2011b. Investing Across Borders: Indicators of Foreign Direct Investment; Ethiopia. Database. World Bank, Washington, DC.

Chapter 6

Skills

Weak entrepreneurial and worker skills are a binding constraint on the ability to improve productivity for small and medium enterprises in several sectors and an important constraint for most firms. By entrepreneurial skills, we mean the managerial and technical skills of the firm's owner or manager. Managerial skills ensure that the production process is efficient and reliable and that the business is well known. Technical skills include the know-how and ability to innovate into new products and to improve the quality or change the features of existing products in response to shifts in market demand. Productivity is also affected by worker skills, especially in the wood sector where the technical skill required for efficient production and high-quality output is greater.

Entrepreneurial Skills

There is considerable heterogeneity in firm performance in Africa, reflecting partly the level of entrepreneurial and management skills and partly the lack of competitive pressure in many countries (Clarke 2011b). Inefficient firms are not driven from the market, and entrepreneurs find entry difficult. This is not unique to Africa. A study finds that one major reason that firms in India do not adopt better management practice and still exist is the limited competition in the domestic market (Bloom and others 2011). A 50 percent tariff on textile imports allows inefficient firms to survive against Chinese imports, resulting in an equilibrium with low productivity, low wages, small firms, and widely dispersed firm productivity (Bloom and others 2011; see also Lewis 2005; Bartelsman and Doms 2000; Hsieh and Klenow 2009).

Gaps between the least and most productive African firms are very large. In Tanzania firms at the twenty-fifth percentile in labor productivity produce about US$1,250 of output for each worker. Firms at the seventy-fifth percentile produce about US$9,050 per worker—about seven times more. In Zambia manufacturing microenterprises at the seventy-fifth percentile produce about US$2,000 of output per worker—about four times more than microenterprises at the twenty-fifth percentile (Clarke 2011b; Conway and Shah 2010). Although the comparisons are for labor productivity, similar dispersions are evident for total factor productivity.

Large variations in productivity within narrowly defined industries arise from multiple factors: limited dispersion of entrepreneurial and technical skills, market segmentation arising from policy interventions (for example, tariffs or entry restrictions) or geography (for example, poor roads), and limited competitive pressure. The consequences are visible in the gap between the free-on-board (f.o.b.) prices for comparable products earned by Ethiopian and Chinese exporters. The f.o.b. price of Chinese-made polo shirts is approximately US$5.38–US$5.80, substantially higher than the f.o.b. price of Ethiopian-made polo shirts, which fetch no more than US$2.95–US$3.41, with the difference reflecting both the physical quality and finish of the product and the capability of the producing firm to provide other characteristics (delivery time, reliability) valued by the buyers. Against 5 percent production waste in best-practice firms in China, best-practice firms in Ethiopia have 10 percent waste. For wooden chair production in Zambia, production waste in converting lumber into a chair is 15–30 percent, compared with 10 percent for Chinese firms. These differences in the production process (not related to costs, inputs, or output markets) can be attributed to weak entrepreneurial management and technical skills.

Low Level of Education

Our survey of small firms found big cross-country differences in the education of firm owners (figure 6.1). In Tanzania, 70 percent of the owners have at most six years of education—that is, they have completed primary school or less. But in Vietnam fewer than 5 percent of owners have only a primary education. In China and Vietnam, nearly 90 percent of owners have more than some secondary education, but in Tanzania, only 20 percent.

Possible Solutions from Asia

These variations reflect national differences in access to education. In China, for example, the expansion of education meant that by around 1990 nearly all school-age children completed primary school; by 2000, nearly all young Chinese completed junior high school. Recent years have brought rapid expansion in the proportion of youths attending high school. While the educational qualifications of African workers will not match China's achievements for some decades, productivity in light manufacturing can be raised in the short term by increasing the skills of entrepreneurs (through managerial training and technical assistance) and facilitating the positive spillovers of establishing a competitive manufacturing sector (through support to first movers and foreign direct investment). In the two Asian countries, most respondents list family members and relatives as contributors of ideas, technical expertise, and financing. The proportions are much smaller in Africa, especially for technical expertise and financing.

Figure 6.1 Level of Education of Firm Owners in the Five Countries

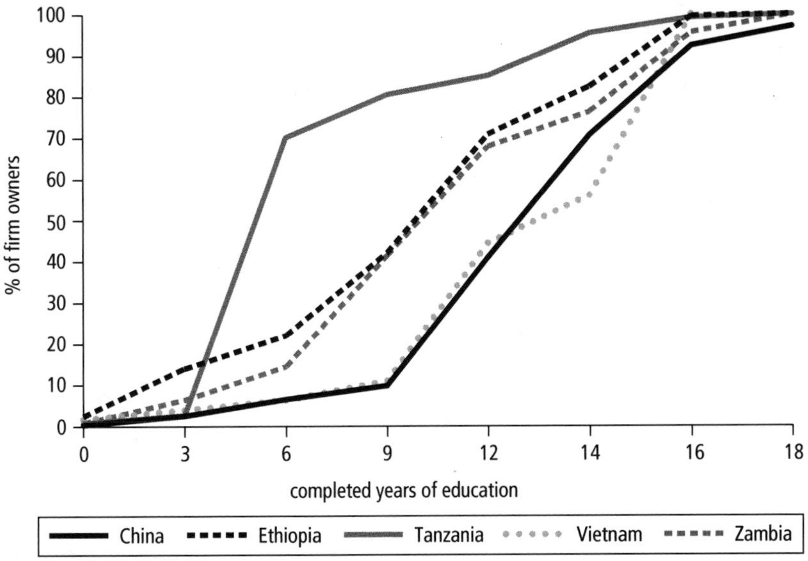

Source: Fafchamps and Quinn 2011.

Does this mean that African entrepreneurs find the ideas, expertise, and funding they need elsewhere? No. Respondents in the three African countries are much more likely to report that no one other than themselves contributed to the creation of the firm. Furthermore, respondents in Asian firms are also more likely to list other sources of help, particularly among Chinese respondents who are much more likely to list business acquaintances, experts, consultants, clients, employees, and equipment suppliers as sources of assistance at start-up.

Reflecting China's legacy of institutions developed during several decades of Soviet-inspired economic planning, government agencies and research institutions provide more technical expertise in China than in the other countries; following more than three decades of incoming foreign investments, Chinese respondents are also more likely than firms in other countries to benefit from the technical expertise of foreign joint ventures. Customers tend to be cited more often as a source of technical expertise in China and Vietnam than in the three African countries. Even sharper differences emerge for experts and consultants, cited by 28 percent of Chinese respondents but by only a few respondents in the other countries. The pattern is similar for research institutions, cited by 11 percent of Chinese respondents but hardly anyone in the other four countries.

Finally, suppliers of equipment and raw materials are cited as sources of advice in China and Vietnam, but not in Africa. By contrast, competitors are cited by numerous firms in Africa, but by fewer firms in Vietnam and China. The picture that emerges is that firms in the two Asian countries, particularly in China, have access to a much wider range of information sources for the technical expertise needed to develop new products.

From this, we can infer that firms in Sub-Saharan Africa are less able to take advantage of links to globally competitive suppliers, customers, and competitors than their Chinese counterparts. Given the same individual skill set of an entrepreneur in China and one in Zambia, the one in China has the advantage of proximity to a competitive local manufacturing sector. The access of entrepreneurs to market information, new technology, improved inputs, production practices, and global buyers is made easier by proximity to globally competitive firms. One entrepreneur does not have to know everything, but entrepreneurs in Africa have far fewer opportunities than entrepreneurs in China to offset shortcomings by accessing the established networks and institutions of a well-developed manufacturing sector.

Sector-Specific Technical Assistance. Receiving help from family and relatives is associated with a smaller firm at start-up, while receiving help from experts or finance from financial institutions is associated with a larger firm at start-up (table 6.1).

It would be perilous to interpret this relationship as causal—entrepreneurs who seek advice from experts and secure start-up funding from financial institutions may simply be better than the average. The relationship nevertheless suggests that obtaining advice and finance at start-up may lead to good entrepreneurship.

To investigate this idea further, we regress (the logarithm of) current firm size and the firm's employment growth rate on the assistance received at start-up. Assistance at start-up remains a strong predictor of future firm performance:

Table 6.1 Firm Size and Assistance Received at Start of Business

Source of help	Firm size at start		Firm size now		Firm's growth rate	
	Coefficient	t-value	Coefficient	t-value	Coefficient	t-value
Family and relatives	−0.213	−3.08	−0.127	−2.56	0.01	1.04
Businesses, clients, and suppliers	0.065	0.90	0.125	2.42	0.01	1.12
Experts and teachers	0.165	1.73	0.283	4.26	0.03	2.05
Financial institutions	0.292	2.74	0.208	2.96	0.02	1.09
Intercept	2.052	34.77	2.667	62.58	0.07	7.25

Source: Fafchamps and Quinn 2011, table 5, which pools data from firms in China, Vietnam, Ethiopia, Tanzania, and Zambia.

firms that received expert advice at start-up on average grow faster than other firms by 3 percentage points a year. Again, help at start-up is likely correlated with business acumen. China, where expert advice is reported by the largest proportion of respondents, also has the highest manufacturing growth rate.

In China the government does not support new firms, except with land and factory shells, but once the firm is established and doing well, government may provide many services (including streamlined administrative procedures, support for technological upgrading, and access to market information through networking) with a view to guiding the firm and the industry to a nationally competitive level.

In Ethiopia the Ethiopian Leather and Leather Products Technology Institute selected the Indian Footwear Design and Development Institute to provide technical assistance for seven shoe factories in design, technical upgrading, quality assurance, productivity enhancement, and testing. As a result of the technical assistance, cutting at Ramsay Shoe Factory rose from 2,000 pairs a day to 2,400, and defect rates dropped from 3 to 1 percent. In July 2010 Ramsay signed a deal with Geox, an Italian shoe and apparel manufacturer, to produce 60,000 pairs a month to be sold in Geox's outlets in Italy, carrying the label "Made in Ethiopia by Ramsay." Clearly, providing the right technical assistance to owner-managers can have a big impact on factory performance.

Direct Training of Entrepreneurs. For entrepreneurs in developing countries such as those in Sub-Saharan Africa, it may not be clear whether managerial training can contribute to productivity, where it can be found, and whether it is affordable. In order to address these issues this study has carried out Kaizen randomized controlled experiments at four industrial clusters in three countries: a garment-textile cluster in Dar es Salaam, Tanzania, a metalwork cluster in Addis Ababa, Ethiopia, a rolled steel cluster in Bac Ninh, Vietnam, and a knitwear cluster in Hatay, Vietnam. In these experiments, two types of management training programs for entrepreneurs were provided (on site and in the classroom), and their impacts on business practices and performance were evaluated based on the data collected from participants and nonparticipants before and after each type of training program. The enterprises were randomly invited and could decide whether to participate or not in any of the training programs.

The findings show that, in the countries studied, standard business practices such as book and inventory record keeping are generally not implemented. Moreover, the number of enterprises that declined to participate in the program was especially high in Ethiopia and Vietnam, suggesting the low expected value from the program, presumably due to a lack of prior information about such programs and their returns. Following the Kaizen program, however, the entrepreneurs' willingness to pay for managerial training increased significantly. Improvements in marketing, quality inspection, record keeping, and Kaizen practices are evident, although the impact of the program in these areas differed across the countries (see Sonobe, Suzuki, and Otsuka 2011 for the detailed results). In addition, findings from Ethiopia's metal cluster show that more entrepreneurs in the treatment groups planned to make additional investments and upgrade their products or organization in the year after initial training than those in the control group. This suggests that the training could also contribute to enterprise growth. These findings lend strong support to the view that small businesses in developing countries, with few exceptions, do not appreciate the potential for managerial improvements to increase profitability and therefore fail to capture gains arising from more efficient management. Accordingly, policy interventions targeted at addressing this information asymmetry and providing public resources for awareness and skills development may be appropriate in supporting manufacturing competitiveness in Sub-Saharan Africa.

The discovery that African manufacturing entrepreneurs often lack access to sources of information that are widely available elsewhere, particularly in China, and may fail to appreciate the benefits linked to the information that is available provides a rationale for governments to go beyond the universally beneficial policy of expanding access to basic education. While fostering full and fair competition, governments should create, fund, and promote programs that are designed specifically to improve the managerial and technical skills of actual and potential entrepreneurs, starting with the high-potential subsectors, including the informal sector.

As the Kaizen study shows, there are two reasons why market forces alone will not allocate optimal resources to create and introduce knowledge that entrepreneurs need in the informal sector (Sonobe, Suzuki, and Otsuka 2011). First, information spillovers and imitation are widespread, with the proximity of products of rival enterprises and the poaching of inputs and technical information from neighbors. Second, entrepreneurs may not know the value of training. As a result, the social return to creating and introducing new knowledge, both technical and managerial, exceeds the private return, resulting in socially suboptimal investment in new, innovative knowledge.

For an enterprise to grow, its manager must be an entrepreneur who constantly strives for innovations. To become a dynamic entrepreneur, the manager must invest in his or her managerial human capital, but usually lacks the financial resources to do so. Even if the manager has resources, he or she may not know where to invest or what to learn. Indeed, the majority of small entrepreneurs in Sub-Saharan Africa operate their businesses without realizing whether they are making profits or losses because they do not keep records of the costs and revenues for their various production activities.

Even a three-week training program can improve entrepreneurs' management practices. The Kaizen program has three modules: marketing and business strategy, production and quality management (including a brief introduction to workplace housekeeping techniques and other Kaizen activities), and business record keeping. Our experience shows that even these brief programs awaken entrepreneurs to the benefits of deepening their managerial expertise, with the result that their willingness to pay for further training increases substantially.

Should Kaizen training be the same for owner-managers of small and large firms? If the training is done in a classroom, the same program will do. If the training is done on site, different approaches have to be taken depending on the firm size (as well as the industry of the firm). In the case of large firms, on-site training is given to a small group or section (5–20 participants). A related question is whether the effect of the training varies with firm size. One of the favorable effects of Kaizen training is to improve communication and coordination among sections within large firms. Two-person firms may experience fewer communication or coordination problems than larger firms, so this effect may be smaller.[1]

Generation and Diffusion of Information. In some cases, African governments should generate and diffuse strategic information on market opportunities in those sectors with potential, yet the private sector is trapped in low growth due to information asymmetries. The information should be specific, timely, verifiable, and targeted to the firms most likely to invest, especially potential strategic first movers. Besides knowledge of market opportunities, the private sector also needs information on production techniques and on new sources and new prices of inputs and machines following the removal of import duties and the reform of trade logistics. With advances in Internet accessibility, much of this information can be collected and disseminated at low cost.

Another novel way to leverage private initiatives with limited public sector resources is to create a "market for intervention."[2] The government could announce that it wants to help firms to improve what they already do by inviting potential beneficiaries to come up with plans and by helping to finance part of the costs—for example, for training workers or conducting market surveys or feasibility studies. The government could also negotiate something similar to an "advance market commitment" by prominent international buyers—for example, an agreement with Macy's to buy 100,000 pairs of shoes backed by a government guarantee to pay 50 percent of the cost if the quality standards are not met.

Efforts to Facilitate the Entry of First Movers. First movers in late-mover countries typically face higher costs and risks, especially in Africa, which has limited infrastructure and high regulatory and governance risks. But strategic first movers can catalyze the growth of competitive new industries. The first rose farm in Ethiopia led to the creation of a new industry that has generated more than US$200 million a year in foreign exchange and created 50,000 jobs since 2000 (box 6.1).

BOX 6.1

Ethiopia's Success in Farming Roses

Ethiopia's first rose farm, Golden Roses, was created in 2000. Since then the farm has triggered a competitive rose export industry that involves more than 75 firms, employs more than 50,000 workers, and earns more than US$200 million a year in foreign exchange. The idea came from the father of the farm's owner, Ryaz Shamji, the Indian head of a Ugandan conglomerate, after visiting Ethiopia to assess business opportunities. Favorable soil and climate conditions (warm days and cool nights), competitive fuel and electricity costs, and, above all, competitive air freight costs—which account for more than half of export-related production costs—made rose farming an easy choice, despite the absence of a strong flower industry.

The first challenge was finding 7 hectares of usable land. Because Ethiopia has no land market, doing so took a year and required intervention from high-level authorities, who gave Shamji's farm a 30-year lease on land abandoned by a nongovernmental organization. The second constraint was financing, because private banks were not willing to lend money to a new venture in Ethiopia. The state-owned Ethiopia Development Bank eventually agreed to provide a loan for 30 percent of the project (US$1 million) at an 8 percent interest rate. Shamji would not have proceeded with the investment without this loan. Another concern was ensuring a reliable water supply, so Shamji investigated his options with help from an Israeli company that specializes in irrigation systems. The final major challenge—a lack of specialized managerial capability—was overcome by convincing an Indian from Kenya and an Israeli to move to Ethiopia.

This fair trade–certified farm made a profit almost immediately. In 2002, based on the farm's success, the prime minister agreed to support the industry by facilitating access to land and providing tax incentives, duty-free imports, and long-term financing for up to 70 percent of the initial investment. With this support and the demonstration effect, investors poured in, enabling the government to meet its goal of developing 800 hectares of rose farms by 2007. Since the government announced its support, more than 75 firms have entered the rose industry.

Shamji and other floriculturists suspect that there is large potential for horticulture, but no one knows which fruits are smart investments for the region—peaches, apples, or something else. A feasibility study could cost more than US$80,000 a fruit, and several studies may be needed. With so many eager incumbents and potential entrants, even Ethiopia's pioneer rose farmer is not willing to bear the disadvantages of being the industry's first mover.

This story embodies the highs and lows of first-mover risks, how these risks were reduced by government policies and high-level government interventions, and how a business's continued growth depends on steady reforms.

Source: World Bank team interview.

First movers can provide many benefits, including demonstration effects for other potential entrepreneurs and their financiers, specialized infrastructure that can be shared, skills development, policy reforms, supplier industries, and improvements in the country's reputation. The first producer of corrugated tin roofs in Zambia attracted a dozen followers, largely replacing imports. The first rose farm in Ethiopia helped to create a land market as farmers realized that they could double their income by leasing their land to rose farms while also working for them. Cadot and others (2011) show that survival rates improve significantly with the number of firms exporting the same product to the same destination. The analysis provides strong evidence suggesting that this is due to information spillovers (for example, increased confidence of importers and financers). Government support to first movers is thus justified by the need to mitigate the higher costs and risks faced by new entrants and by the positive externalities.

Government support to first movers is thus justified by the need to share the risk of investment and uncertainty by reducing the high costs of entry and fostering the positive externalities associated with new firm entry. But such support should be provided only after lifting a critical mass of the most serious constraints so as to remove the need or temptation to provide rents to a few well-connected firms. Furthermore, such support should only be provided on a one-time basis and does not have to cost much. The nature of the market failure should be identified—information asymmetries, free riders, public goods of trade infrastructure, and worker skills—and support should be designed to address that market failure.

Nor does support to first movers have to be targeted exclusively at large firms or at overseas investors: basic knowledge of markets and suppliers may suffice to unleash new industries, as with Zambia's corrugated tin roof industry, which grew rapidly from a single to a dozen firms and began competing with imports. The government could provide a matching grant to share the high risks of first movers by, say, paying for part of a feasibility study.

Worker Skills

Education statistics highlight Africa's current disadvantage in human capital. World Bank data place recent literacy rates of young people ages 15–24 at 45, 75, and 77 percent, respectively, in Ethiopia, Tanzania, and Zambia as opposed to 97 and 99 percent, respectively, in Vietnam and China.[3] Still, we know that China's rapid expansion of manufactured exports began during a period when school attendance and literacy were considerably below their current high levels. Even so, it is clear that the limited dispersion of skills contributes to the current inability of most African light manufacturing to compete effectively in global (and sometimes even in domestic) markets for labor-intensive products.

Table 6.2 Labor Efficiency in the Five Countries

Labor productivity	China	Vietnam	Ethiopia	Tanzania	Zambia
Polo shirts (pieces per employee per day)	18–35	8–14	7–19	5–20	—
Leather loafers (pieces per employee per day)	3–7	1–6	1–7	4–6	—
Wooden chairs (pieces per employee per day)	3–6	1–3	0.2–0.4	0.3–0.7	0.2–0.6
Crown corks (pieces per employee per day × 1,000)	13–25	25–27	10	—	201
Wheat processing (tons per employee per day)	0.2–0.4	0.6–0.8	0.6–1.9	1–22	0.6–1.6
Dairy farming (liters per employee per day)	23–51	2–4	18–71	10–100	19–179

Source: Global Development Solutions 2011.
Note: — = Not available. Crown cork (bottle cap) production in Zambia is fully automated. Figures for wheat processing come from a sample of small firms.

In comparative terms, skills are particularly low in the vast African informal sector, but much less so in the small African formal sector, where productivity in the best firms is sometimes comparable to Asian norms.

Notwithstanding the differences in skills, in all but one subsector (wood products) the efficiency of African workers overlaps the range observed in China and Vietnam (table 6.2). The numbers suggest that low-level skills are sufficient for jobs such as in computer management and technology operations in the apparel industry. Africa's potential can also be inferred by the significant positive impact that targeted technical assistance programs have had on both efficiency and quality (for example, the Ethiopian shoes industry). So Sub-Saharan Africa can be competitive in light industries that do not require semiskilled workers, because there is a plentiful supply of trainable unskilled workers.

To graduate to more complex production such as dress shirts or trousers in the apparel industry, manual or artisanal skills become important. If Sub-Saharan Africa wants to upgrade its industries to more complex products, there is no shortcut to having a labor force with more vocational training. The qualitative interviews and factory visits in China and Vietnam showed that the assembly line operations involving weaving machines to produce sweaters, sports shoes, and toys—or even labeling and packaging products—require workers with more technical skills and higher literacy. Comparative advantage is dynamic: mastering the production of the simplest light manufactures such as T-shirts can open the door to a comparative advantage in more sophisticated light manufactures, but not without a corresponding upgrading of industry skills (Lin 2009, 2010). And because skills upgrading is a public good, there is a role for soft industrial policy (Harrison and Rodríguez-Clare 2010).

Possible Solutions from Asia

Sub-Saharan governments could offer technical assistance to foster industry-specific vocational training for their less-skilled workers. Such training could be offered in industrial clusters for smaller firms and in the workplace or in industrial parks for larger firms. Funding for technical assistance could be leveraged from donors. In the medium term, governments, in partnership with the private sector, could offer publicly funded programs to turn out technicians who can operate and repair simple machines, read instructions, keep systematic records, and use the Internet to communicate and search for information. A good illustration of the potential benefits of such programs is the Penang Skills Development Centre (Malaysia), which offers a variety of sector-specific short- and long-term certificate, diploma, and degree courses for all levels of the workforce.[4]

Dedicated skills development schools for some sectors have had huge pay-offs in countries where they were directed at sectors in which the country had a latent comparative advantage. By focusing on the knowledge and skills for an emerging sector, such schools have rapidly produced a pool of technically trained workers for the sector's takeoff. In Chile the government established the Salmon Technical Institute to train a large pool of skilled workers to scale up the farming of salmon, less known to the smaller fishermen. To exploit India's comparative advantage in fruit cultivation and exports, the government established the Indian Institute of Horticulture Research in 1968. And when it noticed the emerging grape sector, in 1997, it went one step further and established the National Research Center for Grapes. In Malaysia the government set up the Palm Oil Research Institute to foster its palm oil industry. Most of these schools serve multiple purposes—from propagating simple extension practices to conducting high-powered research.

In the early stages of developing its light manufacturing, Sub-Saharan governments could use their latecomer advantage to form twinning arrangements with other developing countries that already have technical schools. The ongoing technical cooperation between India's Footwear Design and Development Institute and Ethiopia's Ministry of Trade and Industry is training Ethiopian leather shoe–producing firms to apply the latest benchmarking techniques to produce globally competitive footwear. This partnership has already delivered substantial payoffs in improving firm productivity, lowering waste and rejection rates, and improving the overall quality of shoes produced.

Notes

1. This may not always be the case. For example, imagine that two workers have to share the same room. Without the training on work layout from Kaizen, one worker may be irritated in looking for a tool, thinking that the roommate has misplaced

it and that this kind of inconvenience would not happen if they did not share the room.

2. We are indebted to Brian Pinto (PRMVP, World Bank) for the suggestions made in this context.

3. See http://data.worldbank.org/indicator/SE.ADT.1524.LT.ZS.

4. See http://www.psdc.org.my/html/default.aspx?ID=9&PID=155.

References

Bartelsman, Eric J., and Marc Doms. 2000. "Understanding Productivity: Lessons from Longitudinal Microdata." *Journal of Economic Literature* 38 (3): 569–94.

Bloom, Nicholas, Ben Eifert, Aprajit Mahajan, David McKenzie, and John Roberts. 2011. "Does Management Matter? Evidence from India." NBER Working Paper 16658, National Bureau of Economic Research, Cambridge, MA.

Cadot, Olivier, Leonardo Iacovone, Denisse Pierola, and Ferdinand Rauch. 2011. "Success and Failures of African Exporters." Policy Research Working Paper 5657, World Bank, Washington, DC.

Clarke George. 2011a. "Wages and Productivity in Manufacturing in Africa: Some Stylized Facts." Background paper (Light Manufacturing in Africa Study). Available online in Volume III at http://econ.worldbank.org/africamanufacturing. World Bank, Washington, DC.

———. 2011b. "Assessing How the Investment Climate Affects Firm Performance in Africa: Evidence from the World Bank's Enterprise Surveys." Background paper (Light Manufacturing in Africa Study). Available online in Volume III at http://econ.worldbank.org/africamanufacturing. World Bank, Washington, DC.

Conway, Patrick, and Manju Shah. 2010. "Incentives, Exports, and International Competitiveness in Sub-Saharan Africa: Lessons from the Apparel Industry." World Bank, Washington, DC.

Fafchamps, Marcel, and Simon Quinn. 2011. "Results from the Quantitative Firm Survey." Background paper (Light Manufacturing in Africa Study). Available online in Volume III at http://econ.worldbank.org/africamanufacturing. World Bank, Washington, DC.

Harrison, Ann, and Andres Rodríguez-Clare. 2010. "Trade, Foreign Investment, and Industrial Policy for Developing Countries." In *Handbook of Development Economics*, Vol. 5 of Dani Rodrik and Mark Rosenzweig eds., 4039–214. Amsterdam: North-Holland.

Hsieh, Chang-Tai, and Peter J. Klenow. 2009. "Misallocation and Manufacturing TFP in China and India." Working Paper 09-04, Center for Economic Studies, U.S. Census Bureau, Washington, DC.

Lewis, William W. 2005. *The Power of Productivity: Wealth, Poverty, and the Threat to Global Stability.* Chicago: University of Chicago Press.

Lin, Justin. 2009. *Economic Development and Transition: Thought, Strategy, and Viability.* Cambridge: Cambridge University Press.

———. 2010. "New Structural Economics: A Framework for Rethinking Development." Policy Research Working Paper 5197, World Bank, Washington, DC.

Sonobe, Tetsushi, Aya Suzuki, and Keijiro Otsuka. 2011. "Kaizen for Managerial Skills Improvement in Small and Medium Enterprises: An Impact Evaluation Study." Background paper (Light Manufacturing in Africa Study). Available online as Volume IV at http://econ.worldbank.org/africamanufacturing. World Bank, Washington, DC.

Chapter 7

Implementation

Policy interventions should begin with pilot case studies and be continually revised and updated as the situation evolves. In addition, implementation should be decentralized, to be closer to firms, increase accountability, and foster competition between local governments. As discussed in chapter 1, the key to selective intervention is to focus on policies that reinforce the country's latent comparative advantage, particularly in sectors where it is possible to identify a short and clear path toward making producers competitive in domestic and overseas product markets. Ideally, the size of government intervention to overcome constraints and information failures will be small, and the duration of such interventions will be brief, allowing them to be scaled back as initial success attracts rising flows of private resources into newly competitive sectors.

Not all efforts to support the chosen industries will succeed. A lesson from East Asia is that the government should be willing to abandon failing policies, so it is important to keep the pilots small. The priorities and sequencing of policy reforms and interventions should satisfy three criteria. First, these measures should focus on sectors and subsectors demonstrating the most potential for comparative advantage and job growth. Second, they should be the most cost-effective in the short and long runs, with the least fiscal impact. Third, implementation capacity, governance, and the political economy of policy reforms should be thoroughly assessed and used as a guide.

The report highlights the importance of industry-specific issues (for example, disease affecting the quality of leather, import tariffs on steel, and price caps on agricultural products) in addition to economywide issues (for example, poor access to industrial land, lack of technical skills, and poor trade logistics). This raises the issue of how to prioritize and package policy reforms within and across sectors given that the government cannot tackle all issues at once. But the economywide and the industry-specific reforms need not be mutually exclusive. They can be complementary, and both types are needed to move the economy forward. Priorities have to be set on the basis of cost-benefit analysis, and in cases where hard choices have to be made, the government will need to prioritize its actions with regard to the locations and industries of highest potential, taking into account the political cost of reforms. Such an approach would also facilitate expanding the reforms through demonstration effects. This is what China did in the case of land reforms (initiated in the Shenzhen Special Economic Zone) and Mauritius did in the case of labor reforms (first limited to exporters). Of course, such deliberate targeting of reforms may be prone to mistakes and capture—hence the importance of putting in place a transparent and professional process to carry them out, together with the political courage to correct mistakes.

Competition

Ongoing patterns of development in light manufacturing in China and Vietnam show that competition is the most critical aspect of competitiveness. In China, government policies to facilitate growth in light manufacturing are designed for selected industries rather than for selected firms and allow open competition. Indeed, the Chinese government has leveled the playing field for even small firms to compete in many subsectors of light industry, and the Vietnamese government is trying to do so. The policy approach in this study proposes active government support for light manufacturing sectors (as in China) and, in the case of first movers, even for specific firms (as in Ethiopia's new rose plantations) through technical assistance, rollback of institutional obstacles (for example, surrounding land acquisition), development of input industries, and creation of industrial parks. Policy and implementation should ensure that the beneficiaries of government policy are determined by market forces and not by the special interests of government officials or rent-seeking firms. In many countries, corruption poses a risk to the government's ability to play a positive role in promoting competitiveness and growth. In China the interests of government officials and society were aligned such that officials are keen to expand investment and growth.

In addition to microeconomic interventions of the sort proposed in this report, governments can facilitate robust private sector growth by maintaining macroeconomic stability and implementing sensible long-term policies for managing important natural resources. African governments have indeed recorded marked improvement in macro-policies during the past decade; achievements include lower inflation, reduced fiscal and trade deficits, and, partly as a result, higher growth of gross domestic product. Ethiopia's efforts to tap the hydro-power potential of its rivers represent an initiative in the resource management field that, if successful, has the potential to reverse long-standing shortages of electricity that have undercut past industrialization efforts. The priorities and sequencing of policy interventions should follow four criteria. First, policy interventions should be undertaken only if a market failure (existence of a pure public good, externalities, noncompetitive markets as a result of policy distortions, information asymmetries, or coordination problems) prevents the private market from adequately playing its role. Second, these interventions should focus on sectors and subsectors that demonstrate the most potential for comparative advantage and job growth. Third, they should be cost-effective in the short and long runs, with limited fiscal impact. Fourth, implementation capacity and implications for governance and the political economy of reforms should be assessed thoroughly.

Public-Private Partnerships

In China (and less so in Vietnam) the public sector has overcome its initial hostility and, despite remaining obstacles associated with property rights, access to finance, rule of law, and so on, has demonstrated growing capacity to support and cooperate with private business, especially at the local level and in sectors like light industry that do not occupy a prominent position in official development plans. In Africa, by contrast, an adversarial relationship often exists between the public and private sectors. Public entities view private firms as a source of rents or as vicious, corrupt profit seekers bent on exploiting the nation's resources. Private firms see the public sector as a source of rent seeking or trouble making. Bridging this gap will take some time.

In the early years of industrial development the Chinese government built a reasonable track record in providing macroeconomic stability and gradually dismantled elements of the former planned economy that had prevented the establishment of openly private enterprises and hampered the expansion of small and medium enterprises in the public sector. Chinese governments provide a wide range of fiscal incentives in the form of tax holidays for newly established companies in industrial zones. They refund value added taxes to exporters. They clear land for development of the zones and engage private developers to provide the facilities and services. They also help industry by cultivating networks and associations.

In recent years rising costs for labor, land, energy, and materials, limited land, and increasing environmental requirements have made it harder for local governments to provide these incentives. But the ongoing commitment of Chinese local governments to fueling local economic growth, which is rooted in the benefits that strong growth provides to local officials in the form of improved prospects for promotion as well as personal emoluments, reflects the success of China's current system in broadly aligning the interests of the center, local administrations, the business sector, and the populace at-large, all of whom have shared in the benefits of sustained growth.

High-Level Government Commitment

For the proposed agenda to attract both domestic and foreign investors to the light manufacturing sectors, private investors contemplating risky investments need convincing assurances from the highest level of government regarding the stability of the supportive policy agenda and the regime's determination to avoid the sort of expropriation that has stripped private firms of carefully built assets and opportunities. Appointing a dedicated high-level implementation team with direct access to top national government leaders will help to reassure investors and simultaneously increase the chances of developing and implementing the proposed reform program in an effective and timely fashion.

The Role of Development Partners

While the policy measures proposed in this report are well within the implementation capacity of African countries such as Ethiopia, Tanzania, and Zambia, they may be taxing for others, especially countries that have recently emerged from conflicts. Development partners (nongovernmental organizations, trade and business associations, multilateral and bilateral donors) can help to remove bottlenecks in specific subsectors. In most cases, multilateral or bilateral donor financing of value chain analyses and feasibility studies could reduce the risks to first movers.

Governance and the Political Economy of Reforms

Despite the limited capacity and weak governance of some Sub-Saharan African countries, the approach proposed will not overburden the administrative capabilities of the target nations for several reasons. First, the magnitude of policy intervention is limited, so any rents generated are also likely to be limited. Second, the proposed policy reforms are based on the latent comparative advantage of the target economies, which means that small, temporary support may attract many new entrants (as in Ethiopia's rose export sector), resulting in intensified competition and obviating the need for additional support. Third, the policy interventions themselves are likely to increase competition (for example, by reducing entry costs and risks) as well as the capacity of firms to compete—the opposite of what the past failed industrial policies did in Africa (such as providing protection and subsidies to a few companies in industries in which the country had no comparative advantage). Finally, the presence of economic expertise at the highest levels of government can surely help them to expedite implementation of the policies proposed in this report and to respond promptly and effectively to difficulties as they arise, as is the case in Ethiopia.

Part 3

Identifying the Potential and Easing the Constraints

Chapter 8

Ethiopia as Exemplar

Ethiopia exemplifies how the various analytical tools can help to identify and resolve the most important constraints in a country by differentiating subsectors and firm sizes (box 8.1). Three main results emerge from the analysis:

- Ethiopia has the potential to become globally competitive in large segments of light manufacturing (apparel, leather products, and agribusiness) and, if successful, could create millions of productive jobs in the process by leveraging its labor cost advantage (low wages combined with high labor productivity in good-practice firms) and comparative advantage in the natural resource industries (agriculture, livestock, and forestry). Other developing countries such as Vietnam managed to create millions of jobs in light manufacturing by leveraging similar advantages and by attracting leading investors into the countries.

- Issues in key input industries (such as lack of quality leather and high cost of wheat and wood) collectively are the main constraints on Ethiopia's competitiveness in light manufacturing because input costs represent more than 70 percent of the production cost in these industries. The other important constraints are poor trade logistics (critical for apparel) and poor access to industrial land, finance, and skills, which particularly affect smaller firms.

- Implementing a short list of specific policy interventions would go a long way toward addressing the main constraints. Specific interventions that could have a short-term impact include permitting unlimited duty-free imports of leather and fabrics, reducing the cost of trade finance, and liberalizing domestic markets for agricultural inputs and outputs. Other potentially fruitful interventions include removing import tariffs on additional light manufacturing inputs, developing plug-and-play industrial parks, improving customs procedures, reducing transport costs, developing collateral markets (machines, land, and cattle), and deploying targeted technical training programs. Once policy makers decide to embark on a substantial program of such interventions, the government should actively identify and encourage prospective lead investors to examine emergent project opportunities in Ethiopia's light manufacturing industries. The government should also invite leading investors to pursue opportunities in key input industries such as agriculture, livestock, and wood plantations, as well as downstream manufacturing sectors like leather products, apparel, and agroprocessing.

Our review focuses on five light manufacturing subsectors. What follows summarizes our findings with regard to development opportunities, existing constraints, and proposed pathways to relaxing those constraints. We then synthesize the results across all subsectors and discuss how a reform program could address institutional, fiscal, and political economy issues that currently limit the prospects for expanding production, employment, and exports in all segments of light manufacturing.

Ethiopia navigated the global economic crisis in 2008–09 better than many other developing countries, encountering only modest declines in exports, remittances, and foreign investments, which have since recovered beyond their pre-crisis levels. Growth in exports and earnings, in conjunction with a relative slowdown in imports, has enabled the buildup of foreign exchange reserves. Overall inflation has dropped to single digits, mainly due to declining food prices, and growth has remained strong at about 8–9 percent a year since 2009. The Ethiopian government is committed to achieving continued growth within a stable macroeconomic framework, in the context of the new five-year development plan (Ministry of Finance and Economic Development 2010). The strategic pillars for the plan include sustaining rapid economic growth through investment in agriculture and infrastructure, promotion of industrialization, enhancement of social development, and strengthening of governance and the role of youth and women.

Although each country has a unique economy, and several aspects of Ethiopia's political environment, governance, and institutions make it a special case in Sub-Saharan Africa, sufficient common factors exist to suggest that Ethiopia is a good exemplar for a fairly large group of African countries. Ethiopia is endowed with a large number of natural resources that can provide valuable inputs for light manufacturing industries that serve both domestic and export markets. Among its abundant resources are a large cattle population whose skins and hides can be processed into leather and its products; forests whose wood can be used in the furniture industry; cotton that can be used in the garments industry; and agricultural land and lakes that can provide inputs for agroprocessing industries. Ethiopia has abundant labor whose low costs offer it a comparative advantage in less-skilled labor-intensive sectors such as light manufacturing. Several negative factors are also common to Ethiopia and other low-wage African countries, such as shortage of industrial land, poor trade logistics, particularly in landlocked countries, and access to finance.

This report does not claim that these Ethiopia-specific findings can be generalized to all Sub-Saharan African countries, although, where commonalities

BOX 8.1

Analytical Tools

The study draws on five analytical tools applied in 2010–11:

- New research work based on the Enterprise Surveys
- Qualitative interviews with about 300 enterprises (both formal and informal of all sizes) by the study team in Ethiopia, Tanzania, and Zambia and in China and Vietnam, based on a questionnaire designed by Professor John Sutton (London School of Economics)
- Quantitative interviews of about 1,400 enterprises (both formal and informal of all sizes) by the Centre for the Study of African Economies at Oxford University in the same countries, based on the questionnaire designed by Professor Marcel Fafchamps and Dr. Simon Quinn (Oxford University)
- Comparative value chain and feasibility analysis based on in-depth interviews of about 300 formal medium enterprises in the same five countries, conducted by the consulting firm Global Development Solutions, Inc.
- A Kaizen study on the impact of managerial training for owners of small and medium enterprises. This training, delivered to about 550 entrepreneurs in Ethiopia, Tanzania, and Vietnam, was led by Japanese researchers from the Foundation for Advanced Studies on International Development and the National Graduate Institute for Policy Studies.

are found across the three study countries, they are discussed in part II of this volume (chapters 2 through 7). But the new analytical approach applied to Ethiopia can be replicated for other African countries to derive specific diagnoses and propose solutions tailored to country circumstances. Detailed policy recommendations for Tanzania and Zambia will be available shortly. The replication of this methodology beyond our countries of focus will enable a rich analysis of the constraints on light manufacturing in other Sub-Saharan African countries and provide concrete policy recommendations to facilitate the growth of this sector across the region. This analytical approach is what distinguishes this study from others (see box 8.1).

Apparel: Solving Trade Logistics Issues

This section discusses Ethiopia's potential to raise output, create employment, and become globally competitive in the apparel sector, provided that the key constraints to competitiveness are addressed.

Ethiopia's Potential

Ethiopia's past performance in apparel has been poor (US$8 million in exports in 2009, a thousandth of the US$8 billion in exports for Vietnam, which has a similar population). The sector employs only 9,000 workers (compared with 1 million in Vietnam), most of them producing low-quality products for the domestic market in small, low-productivity firms. Only a handful of large firms (employing about 1,000 workers) have managed to export apparel—mostly non-time-sensitive low-value items. But the detailed comparative value chain analysis (Global Development Solutions 2011) shows that Ethiopia has the potential to become globally competitive in apparel, thanks to the following:

- A large underemployed workforce and a significant and growing labor cost advantage (wages are a fifth of China's and half of Vietnam's; consideration of nonwage labor costs would widen Ethiopia's potential advantage)

- Trainable workers, with Ethiopian workers in the few exporting firms producing as many polo shirts of the same quality per day as workers in Vietnam (half of China's) using similar technology, giving a 5 percent production cost advantage over both Vietnam and China (Global Development Solutions 2011)

- Access to a state-of-the-art container port (Djibouti), with a favorable geographic position

- Potential to develop a competitive cotton or textiles industry due to good climatic and soil conditions together with cheap hydro-energy (under construction), saving both on transportation costs and on delivery times

- Duty-free access to the European Union (EU) and U.S. markets (as with the African Growth and Opportunity Act).

With a population similar to Vietnam's, Ethiopia has the potential to create as many jobs in this sector as Vietnam, provided several constraints are addressed.

Main Constraints to Competitiveness

The binding constraint on Ethiopia's global competitiveness in apparel is poor trade logistics, which nearly eliminate the country's wage advantage (figure 8.1). Poor trade logistics also eliminate Ethiopia from the higher-value time-sensitive segments of the market: elapsed time between order and delivery exceeds 90 days and remains unpredictable. The competitiveness penalty resulting from poor trade logistics is compounded by the absence of competitive input industries (such as textiles), forcing firms to import inputs at high costs and with long, unpredictable delays. Poor trade logistics are more binding for large exporting firms. But as small and medium enterprises become more competitive and link with large firms, they too will be constrained by trade logistics.

Figure 8.1 Costs to Produce a Polo Shirt in Ethiopia Compared with Costs in China

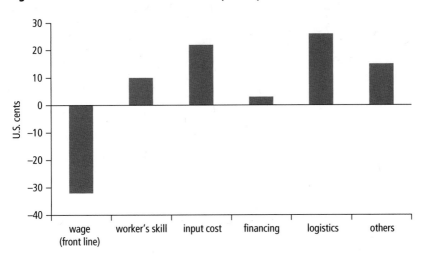

Source: Global Development Solutions 2011.
Note: Here, worker skills cover labor efficiency.

Poor trade logistics in Ethiopia can be attributed to two factors: high handling and financing costs and physical distance from production centers to the
Djibouti port:

- The high handling and financing costs have four components. Charges by
  commercial banks are steep for trade-related financial services such as letters of credit: letters of credit cost 3 percent for imports and 2 percent for
  exports, compared with less than 1 percent for firms in China and Vietnam.
  This imposes a 2 percent production cost penalty on exporters. The National
  Bank of Ethiopia charges a 1.5 percent foreign exchange commission for the
  dollars needed to import the inputs—adding a further 1 percent production
  cost penalty for exporters. Waiting for the bank's authorization creates long
  and uncertain delays (up to six months when foreign exchange is scarce).
  Dealing with customs adds a 0.5 percent production cost penalty; in addition, customs procedures take 25 days longer in Ethiopia than in China. Port,
  handling, and customs fees are higher in Ethiopia. It costs US$600 to handle
  an imported 20-foot container in Djibouti, compared with US$80 in China.

- Ethiopia's manufacturing is centered in Addis Ababa, requiring the transport
  of inputs and outputs between Addis Ababa and Djibouti port 800 kilometers
  away. In addition to the long distance, other factors compound the problem.
  First, the lack of competition in trucking and the absence of railways, coupled
  with high fuel taxes, mean very high transportation costs. Second, it costs

60 percent more to ship to the United States from Djibouti than from China. Shipping costs from Djibouti to Europe are about the same as from China, despite Ethiopia's relative proximity to European ports, due to less traffic in Djibouti and competition issues in shipping. Together, these two factors add up to a 3 percent production cost disadvantage and a six-day delivery lag (compared with China) for shipments bound for European ports.

Other important constraints include the following:

- *Lack of domestic input industries, particularly textiles.* Ethiopia has an ample supply of cotton, but the spinning and ginning industries are weak (a legacy of state ownership of textile mills), and most apparel accessories (such as buttons) still need to be imported. One firm with foreign investors has invested heavily in transforming cotton into fabric, to serve its apparel production. Firms in China and Vietnam do not need to make such investments because either the domestic supply is ample or importing inputs is easy.

- *Difficulties in accessing industrial land and finance for small firms.* Larger exporters have preferred access to both. Small firms have one or two rooms, with a shop front and workshop. Their scope for expansion is limited by difficulties in getting finance to buy land and buildings and to upgrade machinery (Fafchamps and Quinn 2011). This is aggravated by the fragmented land market, with inflated prices and long delays in transactions.

- *Weak entrepreneurial and worker skills, especially among smaller firms.* Weak skills prevent domestic firms from plugging into the global supply chain. The apparel market demands rapidly changing designs, so apparel producers need to be well linked to buyers and able to adapt production designs to meet the quality and designs demanded by the global market. The quantitative survey shows that small entrepreneurs in Ethiopia have less access to skills and information than their Asian counterparts (Fafchamps and Quinn 2011). Good-practice entrepreneurs in both Asia and Ethiopia have demonstrated that low-skilled workers could achieve high productivity with proper incentives and a few weeks of on-the-job training.

Recommendations to Unleash Ethiopia's Potential

Addressing a critical mass of trade logistics issues would put Ethiopia in a position to follow the examples of China and Vietnam by attracting overseas investors to lead the industry's development and provide training and experience for prospective domestic followers. Taking advantage of Ethiopia's high-quality cotton and expected cheap hydro-electricity to develop a competitive textile industry would further enhance the prospects for building competitive strength in apparel.

In the short term, bringing the cost of letters of credit to the level in China and Vietnam would reduce exporters' free-on-board (f.o.b.) production costs by more than 3 percent. Eliminating the foreign exchange fee (1.5 percent) charged by the National Bank of Ethiopia and guaranteeing the immediate availability of the foreign exchange for all apparel producers would reduce f.o.b. production costs for exporters by 1 percent and considerably reduce the length and uncertainty of deliveries. Establishing a "green channel" for apparel at customs would reduce f.o.b. production costs by 0.5 percent and slash customs delays from 30 days to five, the same as in Hong Kong SAR, China. Negotiating lower port and handling fees as part of a strategic partnership with Djibouti could reduce f.o.b. production costs by 1 percent.

For the 800-kilometer stretch between Addis Ababa and Djibouti port, a multimodal transport arrangement combining trucking, rail, air, and shipping would improve connectivity and increase competition. Rehabilitating the railroads from Addis Ababa to Djibouti could further reduce transportation costs and delays. Competition in trucking could be fostered by opening it to new entrants, including foreign companies, and creating a level playing field for domestic companies. Lowering the fuel tax and import tariffs on trucks and spare parts would reduce f.o.b. production costs for manufactured goods by around 0.5 percent. Developing a plug-and-play industrial park in the vicinity of Djibouti (for example, at Dire Dawa, where the climate is not too hot) would reduce f.o.b. production costs by 2 percent and shave six days from the time to deliver the products. Both China and Vietnam started their apparel production industries by setting up industrial zones next to world-class ports.

To deal with other less binding constraints, Ethiopia could facilitate access to inputs (beyond improving trade logistics) through three sets of measures:

- *Eliminate all import tariffs on apparel inputs.* Import tariffs on apparel inputs now range from 10 to 35 percent. Duty-free access to inputs is limited to exporters, with duty levied on any final goods not exported. Eliminating the duty would enable exporters to resell their material waste (reducing production costs by 1 percent) and facilitate links between (large) exporters and (small) domestic producers. This change would increase productivity and output growth among small players and give exporters greater flexibility to meet large orders. Such a measure would also facilitate implementation of a "green" customs channel for apparel—another inexpensive and beneficial reform.

- *Support competitive input industries, particularly textiles.* Preliminary evidence points to the possibility that Ethiopia could develop a competitive textile industry by taking advantage of its favorable climatic and soil conditions for cotton production and expected competitive sources of hydroenergy (detailed feasibility study to be conducted). Eliminating imports of fabric would reduce production costs by 3 percent and delivery delays by almost half. The current policy of banning exports of cotton to ensure that textile mills are supplied may be counterproductive as it lowers demand and hence the returns to cotton growing, thus discouraging investment in a sector that could contribute to large-scale future employment growth in textiles and especially in apparel.

- *Develop a plug-and-play industrial park in Addis Ababa and conduct a feasibility study for one closer to Djibouti.* As demonstrated in China, such parks can solve several constraints simultaneously by providing firms with affordable access to industrial land, standardized factory shells, and worker housing, as well as training facilities and one-stop shops for business regulations. They considerably reduce financing costs and risks for the better small firms, allowing them to grow into medium enterprises before they are sufficiently large and financially secure enough to obtain bank loans. By creating industrial parts, China has avoided the "missing middle" problem.

Once a program to ameliorate a critical mass of trade logistics issues takes shape, government agencies should promote Ethiopia globally as a destination for investment among leading apparel firms. In addition to providing immediate infusions of capital, foreign exchange, and technical, managerial, or marketing expertise, substantial foreign investment can trigger major externalities, as Daewoo did for Bangladesh by training a new generation of apparel entrepreneurs.

Leather Products: Dealing with Shortages of Quality Leather

The leather subsector has even greater potential for production, exports, and employment for Ethiopia than the apparel sector. Facilitating the import of quality processed leather in the short term while facilitating the development of a competitive leather supply chain would position Ethiopia to become one of the leading global centers for producing quality leather goods.

Ethiopia's Potential

Ethiopia's current performance is as poor in the leather product industry as in apparel. Exports in 2010 were US$8 million, compared with US$2.3 billion for Vietnam. Employment is only 8,000, mostly in small firms producing low-quality items for the domestic market, compared with 600,000 in Vietnam.

Ethiopia's potential in the leather product industry stems from the following:

- Ethiopia has a 12 percent production cost advantage over Vietnam and 37 percent over China because this subsector is even more labor intensive than apparel (labor accounts for 40 percent of total production costs in China, compared with 10 percent in apparel). As in apparel, Ethiopia combines very low wages with highly trainable labor, as demonstrated in the larger firms, where Ethiopian workers produce as many shoes of similar quality per day as workers in Vietnam (80 percent of the Chinese level).

- Thanks to good climatic and soil conditions, Ethiopian cattle produce some of the world's best leather. Ethiopia is among Africa's leading cattle-producing countries; the combination of ample supplies of high-quality leather and competitive hydro-energy should enable the development of a world-class leather industry.

- The competitiveness penalty due to poor trade logistics is less serious in leather goods than in apparel. Ethiopia's wage cost advantage exceeds the trade logistics cost penalty, and the subsector is less time sensitive than apparel.

- As with apparel, Ethiopian leather products have duty-free access to the EU and U.S. markets.

Main Constraints to Competitiveness

The most binding constraint is the shortage of quality processed leather. With full liberalization of leather imports, Ethiopia would already be in a position to be globally competitive. Low wages and high labor intensity mean that the 10 percent production cost disadvantage due to poor trade logistics (assuming that leather has to be imported) would not eliminate Ethiopia's production cost advantage (figure 8.2). The additional order-to-delivery delays would also not be a serious threat to competitiveness in this industry, which is much less time sensitive than apparel (fashions in men's leather shoes change slowly).

Figure 8.2 Costs to Produce Leather Shoes in Ethiopia Compared with Costs in China

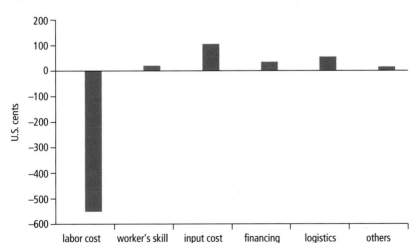

Source: Global Development Solutions 2011.
Note: Here, worker skills cover labor efficiency.

The shortage of quality domestic processed leather has three causes:

- *Poor disease control and weak veterinary services.* As a result of disease, only small pieces of skin are free of blemishes. This is largely the result of failure to control for ectoparasites (a disease that causes skin blemishes) and the domination of the livestock industry by farmers who own only a few animals and sell skins as a side business at best.

- *Lack of quality processing of raw hides and skins.* Ethiopia still suffers from the legacy of government control over tanneries. Recent private investments and technology upgrading offer brighter prospects if the supply of quality skins can be improved. Taxing exports of skins and semiprocessed skins may backfire by artificially reducing domestic prices, thus reducing the incentives for increased private investment along the supply chain.

- *Import bans on processed leather.* The Ethiopian government has effectively banned the import of processed leather through cumbersome administrative requirements. For example, companies willing to import leather said that they have to justify why they cannot source the leather from domestic tanneries.

Other constraints are less important:

- *Poor trade logistics.* Although not a binding constraint, poor logistics result in a 10 percent production cost penalty (6 percent if leather does not need to be imported) for the same reasons as in apparel.

- *Lower product quality due to lack of entrepreneurial managerial skills.* Leather loafers exported from Ethiopia are of a lower quality than those made in China, as demonstrated by the lower f.o.b. price for leather shoe manufacturers in Ethiopia. This difference in quality is in part due to the low managerial and technical expertise of firm owners and managers. The Leather Institute in Ethiopia was established with technical assistance from India to improve managerial skills. In addition, foreign investors have already begun to enter the leather products sector. The success of some firms in entering the Italian market demonstrates the potential for improved productivity and quality to transform Ethiopia's leather sector into a competitive industry capable of generating large increases in employment and exports.

- *Difficulties on the part of smaller firms accessing land, finance, and skills.* These difficulties keep smaller firms in a low-productivity, low-quality trap, unable to upgrade technology or expand production.

Recommendations to Unleash Ethiopia's Potential

Enabling imports of high-quality processed leather would be a quick and easy solution to the acute shortage of leather in the industry. Removing restrictions on exports of leather would increase the incentives to invest in the Ethiopian leather supply chain.

The government should pursue efforts to control cattle disease, a battle that, when won, will also benefit the dairy and meat-processing industries. A U.S. Agency for International Development (USAID) study shows that the infestation rate of ectoparasites could be substantially reduced from 90 to 5 percent with four treatments a year for each animal, costing about US$0.10 for all four treatments (USAID 2008). The total cost for such a program covering the whole country would be less than US$10 million a year, a modest amount in relation to the potential benefits. As discussed in the section on agribusiness, promoting cross-breed cows and enabling the use of cattle as collateral would also increase the incentives and financial means of small farmers to access veterinary services. Facilitating access to rural land for good-practice animal farms would further stimulate the supply of quality leather. This should be done through a fully transparent and inclusive process to ensure that local communities benefit and environmental issues are resolved.

To address the lesser constraints, several measures could be taken:

- *Improve trade logistics.* Improving trade logistics would further increase Ethiopia's competitiveness in leather products by reducing production costs by as much as 5 percent (if leather is imported) and order-to-delivery delays by more than 30 days.

- *Provide technical assistance to improve entrepreneurial skills in product design and factory floor operations.* As the experience of the Ramsay Shoe Factory shows, technical assistance to Ethiopian shoe manufacturers—in this instance, by the Indian Footwear Design and Development Institute—helped to improve labor efficiency as well as the quality and price of exportable shoes.

- *Facilitate the entry of leading investors in leather products.* Interest is already high from both producers (official Chinese policy now encourages domestic companies to expand overseas investments, with leather products now becoming a sunset industry in China's coastal cities) and global buyers (the leading U.S. importer of leather shoes is seeking to diversify away from China). This creates an opportunity to promote Ethiopia as a promising investment location for the global leather industry. To avoid rent-seeking behavior, interventions to attract investment should be available to all firms seeking to enter the industry.

- *Develop plug-and-play industrial parks.* One park should be developed next to Addis Ababa and one not too far from Djibouti (following an in-depth feasibility study) to help to address the missing middle and further improve trade logistics.

Wood Products: Providing Technical Training and Developing Sustainable Wood Plantations

Ethiopia could become competitive in wood products by providing technical training to entrepreneurs and workers, making access to industrial land and finance easier, and by facilitating the development of sustainable wood plantations to take advantage of the country's unique bamboo resources—the biggest bamboo endowment in Africa.

Ethiopia's Potential

Unlike apparel and leather, where the main opportunity is for Ethiopia to develop exports, the immediate opportunity in the wood product industry lies in the fast-growing domestic market for domestic fuel, construction materials, and furniture. The opportunity for job creation is more limited because labor productivity is very low (a fifth of Vietnam's and a tenth of China's on average). In the initial phase of manufacturing growth, this subsector's potential seems to lie in import replacement rather than in exports.

Fueled by strong domestic demand for wood construction materials and furniture, Ethiopia's wood product industry currently employs more than 40,000 workers, mostly in small, low-productivity informal firms producing low-quality products for the domestic market.

Main Constraints to Competitiveness

The cost of producing a wooden chair is more than twice as high in Ethiopia as in China and Vietnam for two main reasons. First, soft wood is much more expensive in Ethiopia than in China and Vietnam—a cubic meter of pine lumber costs US$667 in Ethiopia compared to US$344 in China and US$275 in Tanzania. Second, labor productivity is very low, even in the larger firms. Workers produce 4.5 chairs a day in China, 1.9 in Vietnam, and only 0.3 in Ethiopia (0.5 in Tanzania and 0.4 in Zambia). Lower wages do not compensate for lower productivity, which is caused by the small scale of operations and low skills of managers and workers (the labor cost per chair in Ethiopia is US$10, compared with US$3 in China and Vietnam).

The main constraints are as follows:

- *Lack of organized supplies of domestic Ethiopian wood.* The wood used in Ethiopia is either imported or from high-cost informal sources.
- *Lack of entrepreneurial and worker skills.* The sector is dominated by micro, mostly informal, firms and carpenters producing very low-value products, with a few medium-size companies catering to higher-value niches in the export and domestic markets. For firms of all sizes, poor worker skills are

reflected in labor productivity considerably lower than that of workers in China and Vietnam. Poor management of the production process is also reflected in high material waste compared to China and higher use of consumables such as glues and varnish (table 8.1).

- *Difficulties in access to land and finance.* Several furniture manufacturers in Addis Ababa said they were at maximum capacity in their current premises, but due to the fragmented land market they could not obtain affordable, convenient land to scale up production.

Table 8.1 Benchmarking the Cost of Consumables per Chair in the Five Countries

Country and consumable cost and use	Cost per kilogram (US$)	Grams per chair	Cost per chair (US$)
Ethiopia			
Adhesives, glues	2.59	200	0.52
Varnish, finishing oils	2.22	500	1.11
Other consumables	—	—	2.59
Total	—	—	**4.22**
Tanzania			
Adhesives, glues	3.87	250	0.97
Varnish, finishing oils	2.67	500	1.33
Other consumables	—	—	2.80
Total	—	—	**5.10**
Zambia			
Adhesives, glues	5.85	250	1.46
Varnish, finishing oils	3.47	500	1.74
Other consumables	—	—	1.10
Total	—	—	**4.30**
China			
Adhesives, glues	2.21	44	0.10
Varnish, finishing oils	5.10	50	0.25
Other consumables	—	—	0.16
Total	—	—	**0.51**
Vietnam			
Adhesives, glues	3.24	61	0.20
Varnish, finishing oils	7.54	112	0.84
Other consumables	—	—	0.44
Total	—	—	**1.48**

Source: Global Development Solutions 2011.
Note: — = Not available.

Recommendations to Increase Competitiveness

With sales of wood products oriented primarily toward domestic buyers and employment concentrated within small firms, circumstances are quite different in this sector than in apparel and leather goods. We cannot expect medium- or large-scale, foreign-owned, export-oriented operations to play a lead role in elevating productivity and skills in the short term. Our main recommendations, to be pursued in parallel, are to undertake the following:

- *Facilitate access to land and financing for sustainable private wood plantations of fast-growing species on degraded land close to the main urban centers.* This would improve the quality of wood (such as dried and precut wood) and reduce its cost (via economies of scale and lower transportation costs) so long as it is not overtaxed. Carbon financing should be available for such schemes because this increases the afforested area. China, Vietnam, and Tanzania all have successful wood plantations. Facilitating access to land and finance could encourage similar developments in Ethiopia.

- *Develop plug-and-play industrial parks.* Operating in industrial parks would help the better small and medium enterprises to access utilities, land, finance, and skills, with technical assistance programs for both owner-managers and workers.

- *Improve trade logistics.* Better logistics would reduce the cost of low-value imported wood by 20 percent, but imported wood would still be 50 percent more expensive than competitively produced domestic wood (low-value wood is a very low value-to-weight item). Improved trade logistics would also reduce the cost of other inputs that may not be competitively produced in Ethiopia (such as adhesives, varnish, and finishing oils).

Metal Products: Reducing the Cost of Steel and Providing Technical Training

As in the case of wood products, Ethiopia's competitiveness in metal products is constrained by poor skills and high input costs; steel costs 30 percent more than in China because it is imported and subject to import tariffs. With proven reserves of iron ore deposits and cheap hydro-energy, it might be feasible to develop a competitive steel industry in Ethiopia.

Ethiopia's Potential

Ethiopia exported US$41 million of steel products in 2009, while importing more than US$350 million, an imbalance that points to the possibility that domestic products can at least partially replace imports in metal products, a sector with a fast-growing domestic market. Informal micro firms dominate the metal products industry in Ethiopia. Only one firm produces crown corks (bottle caps); there is no domestic production of padlocks (both items are subjects of the comparative value chain analysis in Global Development Solutions 2011). Ethiopia could develop a competitive domestic industry in metal products:

- As in the apparel and leather subsectors, wages are low and productivity can be high. Despite obsolete equipment, Ethiopian workers achieved half the productivity of Chinese workers in the crown cork industry.
- Many metal products are heavy and bulky, with a low value-to-weight ratio, making imports relatively expensive (even with improved trade logistics).
- Iron ore deposits together with competitive sources of hydro-energy create the possibility that Ethiopia could launch a competitive steel industry (although lack of scale and high cost of capital may stand in the way).

Main Constraints to Competitiveness

The main reason that the production costs of metal products are 20 percent higher in Ethiopia than in China is that the steel in Ethiopia is 30 percent more expensive (tin-free steel costs US$1,414 a ton in Ethiopia, US$1,106 in China). Steel has to be imported at high cost (like wood, it is a low value-to-weight item) and is subject to a 10 percent import tariff.

- *High import tariffs, compounded by poor trade logistics, are the main driver of the high price of imported steel.*

Many of the secondary constraints affecting small and medium enterprises in metal products are similar to those in wood manufacturing:

- *Small and medium enterprises cannot expand because of restrictions in acquiring formal finance and land.* One medium-size manufacturer of metal roofing and construction materials made several attempts to invest in forward and backward integration. Despite securing commitments to invest from foreign partners, the projects failed to take off partly due to difficulties in gaining permission for shared equity with foreign partners. Investors see the potential for growth, but access to finance is a constraint on efforts to upgrade the industry.
- *Lack of entrepreneurial and worker skills is* evident in a 32 percent metal scrap rate, much higher than the 5 percent in other countries. Equipment in Ethiopia is obsolete, reflecting the fact that firm owners and managers are not exposed to global competition (the crown cork producer is the sole supplier to the brewery; see Global Development Solutions 2011).

Recommendations to Increase Competitiveness

Our main recommendations for the metal products sector are the following:

- *Reduce the 10 percent import tariff on steel (with little fiscal impact)*

- *Improve trade logistics* (along the lines discussed above), as this could save another 7 percent on the cost of imported steel
- *Promote the exploitation of iron ore deposits and conduct an in-depth feasibility study* to assess the potential competitiveness of a domestic steel industry.

Agribusiness: Reforming Key Agricultural Input and Output Markets and Facilitating Access to Land and Finance

To unleash the potential of agribusiness, the government should reform agricultural input and output markets, enable the use of land and cattle as collateral, and facilitate access to rural land for investors through a transparent and inclusive process benefiting local communities.

Ethiopia's Potential

As shown by the success of the Ethiopian coffee and cut-flower industries, Ethiopia's agribusiness potential is great for both the domestic and export markets. Vietnam created 450,000 direct jobs in the agribusiness industry, and Ethiopia has the potential to do the same.

The industry's potential stems from the following:

- Low wages combined with good labor productivity result in competitive wheat-milling processing costs (see Global Development Solutions 2011)
- Exceptional and varied climatic and soil conditions for growing a variety of crops, including cotton, wheat, coffee, and fruits, that are suitable for exports, industrial processing, or both
- The second largest livestock population of Africa (behind Sudan)
- Large tracts of unused arable land and low yields on cultivated land
- Proximity to the Gulf, a large importer of food products.

Yet millions of Ethiopians continue to depend on emergency food aid, and Ethiopia imported US$368 million of cereals in 2009. The agribusiness industry is underdeveloped, with only 46,000 workers, mostly in large firms (except in the dairy industry).

Main Constraints to Competitiveness

In the agribusiness sector, as in apparel, leather, and wood products, the main constraint is the high cost and low quality of agricultural inputs—an ironic circumstance given the advantage of favorable soil and climatic conditions. A ton of wheat costs US$330 in Ethiopia, compared with about US$200 in China and Vietnam. The quality of milk is low in Ethiopia (due to poor animal husbandry and transport conditions), and the cost is high (US$0.37 a liter compared with

US$0.13 in New Zealand and US$0.22 in Vietnam). The analysis of the main constraints on Ethiopia's agribusiness industry is based mostly on the detailed comparative value chain analysis of the wheat flour and dairy subindustries (Global Development Solutions 2011).

Various distortions are evident in the agricultural input and output markets:

- Shortages of high-yielding seeds due to inefficiencies in the government-run seed production and distribution systems as well as delays in the certification of imported seeds result in unnecessarily low crop yields and therefore high unit costs, which extinguish the prospects for potentially viable development of downstream processing operations.

- A fixed-price system in the dairy sector removes incentives for dairy farmers to expand production or upgrade technology. The average dairy farm in Ethiopia has fewer than 10 animals, but dairy farms in China and Vietnam have hundreds of cows under a single operation.

- More generally, price caps on key food items may protect low-income consumers in the short run (with the risk of shortages and high prices in the black market), but restrict private investment in the long run.

Access to large areas of land for large-scale farming is restricted due to the confluence of traditional rights with government control of land, so land transactions are lengthy, costly, and risky.

Neither agricultural land nor cattle can be used as collateral for loans, preventing farmers from formal finance to increase productivity and expand production. Only 5 percent of the cows in Ethiopia are high-yield cross-breed cows, which produce much more milk for longer periods (table 8.2). This low yield increases the milking, collecting, husbandry, and overhead costs (including veterinary services, which are a fixed cost per cow). It is not economic to invest in efficient milking equipment and veterinary services for low-yield cows. This leads to a cost penalty of about US$0.10 per liter of milk (a 50 percent increase over the cost of milk in Vietnam). Cross-breed cows can be obtained through artificial insemination or can be bought at a cost of about US$1,000, which represents a capital cost of about US$0.02 per liter (with a 15-year life expectancy). They also require more animal husbandry (cross breeds are less resistant to disease) and much more feed than local breeds.

Table 8.2 Attributes of Local and Cross-Breed Cows

Attribute	Local cow	Cross-breed cow
Fat composition (percent)	4.5–5.0	3.5–3.7
Lactation period (days)	239	300
Milk yield (liters a day)	1.3	12

Source: Global Development Solutions 2011.

Recommendations to Unleash Ethiopia's Potential

The recommended policy interventions aim to create incentives to improve productivity and ensure that farmers have access to technical assistance, land, and finance to make improvements.

- *Review government regulations and direct involvement.* The objective would be to reform or eliminate official mechanisms that stifle incentives to invest in new types of enterprises (large-scale cultivation) or in efforts to upgrade existing operations with improved seeds or cattle. The foregoing review repeatedly highlights government-imposed restrictions on agricultural output and input markets as obstacles to realizing potentially very large increases in production, employment, exports, and the supply of materials to several promising light manufacturing sectors.

- *Facilitate access to agricultural land for good-practice investors.* Good-practice new entrants can bring major positive externalities, as with the cut-flower industry, which took off after the Ethiopian government helped the first rose farm to obtain access to 7 hectares of land. This should be done through a fully transparent and inclusive process to ensure that local communities benefit and environmental issues are addressed.

- *Allow land and cattle to be used as collateral.* Such reforms could be piloted in some key locations first. Promoting cross-breed cows and enabling the use of cattle as collateral would increase the incentives and financial means for farmers to access veterinary services.

- *Establish plug-and-play industrial parks.* The creation of such parks would help leading agribusiness enterprises to enter the industry and better-managed small and medium enterprises to access utilities, land, finance, and skills.

Synthesis across the Five Subsectors in Light Manufacturing

This section summarizes the potential and key constraints of Ethiopia across the five studied subsectors and identifies policy priorities.

Ethiopia's Potential

Ethiopia can become globally competitive in the apparel, leather product, and agribusiness industries and compete with imports in the wood and metal product industries, for five reasons:

- Very low wages combined with high trainability of workers
- Potential access to competitive sources of key inputs

- Access to a state-of-the-art container port in Djibouti
- A large and growing domestic market and proximity to large export markets
- Duty-free access to EU and U.S. markets.

Ethiopia thus shares many characteristics with Vietnam, where millions of productive jobs were created in light manufacturing over the past 20 years.

Main Constraints to Competitiveness

Across the five industries that form the focus of this study, policy issues in input industries, particularly in agriculture, are crucial in limiting Ethiopia's ability to activate its latent competitive advantage in major areas of light manufacturing, notably apparel, leather, wood products, and agribusiness. Poor trade logistics, both within Ethiopia and across its borders, raise the costs of accessing external markets for both imports and exports, exacerbating the substantial negative consequences of insufficient access to inexpensive, good-quality domestic inputs. In addition, difficulties in accessing land, finance, and skills obstruct efforts to create and expand small and medium enterprises in sectors in which Ethiopia enjoys a latent comparative advantage.

Government actions are needed in several areas, either to rectify existing policy distortions or to create public goods needed for private sector growth. Among the former are reduction or elimination of tariffs on inputs such as fabric, leather, and steel; reduction of regulations that limit land acquisition; allowance of undeveloped land, cattle, or equipment to be used as collateral against loans; facilitation of entry by leading foreign companies; removal of restrictions on imported seeds and foreign exchange; removal of restrictions that allow banks (apparently government-owned banks) to overcharge on letters of credit and trade documentation. They also include the removal of price controls on dairy products and breaking up of the monopoly on road transport. Among the latter is strengthening veterinary services such as the control of ectoparasites.

The most important constraints vary by subsector and are often specific to the subsector (table 8.3).

Consistent with the traditional agenda for the investment climate, many of the policy recommendations offered here seek to promote competition and reduce transaction costs (via improved trade logistics and lower import tariffs). The detailed subsector-level diagnostics and cross-country comparisons make it possible to focus on a small number of quite specific policy recommendations for each light manufacturing subsector. Some key recommendations, often overlooked, are related to input industries, such as the need to develop sustainable wood plantations.

Table 8.3 Constraints in Ethiopia, by Importance, Size of Firm, and Sector

		Input industries	Land	Finance	Entrepreneurial skills	Worker skills	Trade logistics
Apparel	Smaller	Important	Critical	Critical	Important	Important	
	Large	Important			Important		Critical
Leather products	Smaller	Critical	Critical	Critical	Important		
	Large	Critical			Important		Important
Wood products	Smaller	Critical	Important	Important	Important	Important	
	Large	Critical	Important	Important	Important	Important	
Metal products	Smaller	Critical	Important	Important	Important	Important	
	Large	Critical	Important	Important	Important	Important	
Agribusiness	Smaller	Critical	Critical	Critical	Important		
	Large	Critical	Critical	Important			

Source: Authors
Note: Blank cells are not a priority.

Reforms in agriculture, where Ethiopia has an untapped comparative advantage, are critical to improving the competitiveness of four light manufacturing subsectors (apparel, leather products, wood products, and agribusiness). The agricultural reform agenda deserves its own in-depth study and discussion, but the necessary reforms fall into two broad categories: liberalizing agricultural input and output markets and facilitating access to rural land for good-practice investors. These important reforms could be initially focused and piloted on a few key products and locations. And facilitating the import of key inputs that cannot be produced competitively in Ethiopia should be addressed by both lowering import tariffs and improving trade logistics.

Developing plug-and-play industrial parks and collateral markets is important for all subsectors. Until now, foreign-run industrial parks have been set up in Ethiopia primarily for the benefit of Chinese firms. But the government has identified five sites to develop as industrial parks. Taking early steps to establish clear objectives for future industrial parks can facilitate the development of a suitable institutional framework that will smooth the path to capitalizing on Ethiopia's latent comparative advantage in light manufacturing.

It is also critical to encourage foreign direct investment (FDI) from leading foreign investors in the subsectors with high potential for exports (apparel, leather products, and agribusiness; see box 8.2). But to attract capable, experienced, and well-financed overseas investors, the government of Ethiopia will have to provide convincing evidence that reforms are under way and that completion of the reform agenda will continue to occupy a leading position in official development efforts.

BOX 8.2

The Role of FDI in Chinese Manufacturing and Apparel Industries

Many studies have documented the significant role of FDI in economic development around the world. The experience of numerous economies—including the United States; Canada; Australia; Taiwan, China; Singapore; and, most recently, China and Vietnam—shows how foreign investment has spurred the growth of output, investment, employment, productivity, and exports. FDI contributes to the structural transformation of host economies, to technology adoption, and to industrial upgrading among domestic firms. FDI enables host nations to gain increased access to world markets for goods, technology, and capital. Among recent studies, Chandra and Kolavalli (2006) provide evidence that attracting FDI has been an important strategy in technological adaptation; Harrison and Rodríguez-Clare (2010) find that trade volumes are highly and positively correlated with foreign investment inflows.

FDI's place, however, is more pronounced in some industries than in others. FDI made particularly large contributions to the recent expansion of both China's and Vietnam's apparel industries. China's "open-door" policy, which marked the beginning of a strong relationship between foreign and domestic investors, helped China to become the top host of foreign capital among developing nations and the biggest exporter of many manufactured products, including apparel. Shenzhen's astonishing transformation from a sleepy village to a central component of China's industrial export explosion epitomizes the developmental potential of foreign investment. In 1980 China selected Shenzhen as one of its first special economic zones. Shenzhen's location at China's border with Hong Kong SAR, China, and close to Taiwan, China, and Macao SAR, China, facilitated FDI from overseas Chinese and provided easy access to the deepwater port in Hong Kong SAR, China. Despite initial difficulties—concerns over rigidities and delays led early investors to minimize their interaction with China's domestic economy—Shenzhen rapidly emerged as a magnet for overseas investment and an export powerhouse in a wide array of light industry trades. Three decades later, Shenzhen remains prominent in both manufacturing and exports, with production migrating rapidly from the initial focus on low-end, labor-intensive goods toward higher value added, technologically sophisticated industries such as information and communications gear.

As of 2010 China's stock of inbound FDI, as tabulated by the United Nations Conference on Trade and Development, amounted to US$578.8 billion, and annual inflow of FDI was US$105.7 billion.[1] Foreign ownership is welcomed in many (but by no means all) manufacturing industries, including apparel FDI clusters along China's east coast, particularly in Shanghai municipality and in Guangdong and Jiangsu provinces (National Bureau of Statistics of China 2011, 256). In the past decade, China has become the largest exporter of textiles and apparel, with 2010 exports of US$206.5 billion accounting for one-third of global trade in textiles and apparel, which reached US$612 billion in value.[2] Between 1995 and 1999, 20 percent of all textile firms in China had some form of FDI. These accounted for only one-fifth of the number of firms but for twice the sales volume of purely domestic firms. They employed roughly as many workers as domestic firms, implying that on average they were larger in size. Similarly, value added per worker in firms with FDI was double the comparable figure for purely domestic producers, in part because FDI firms employed more capital per worker (Hu and Jefferson 2002).

Recent research also shows support for FDI's positive role in the development of China's domestic industries. Du, Harrison, and Jefferson (2011) find that FDI generates positive productivity spillovers through contacts between foreign affiliates and their local clients in upstream or downstream sectors.

There is also evidence that successful industrial policies can attract more FDI, which in turn generates higher growth. Alfaro and Charlton (2007) find that FDI inflows are likely to be higher in targeted sectors and that FDI in these sectors generates higher growth. Du, Harrison, and Jefferson (2011) show that foreign investors targeted via special Chinese tax incentives generated significantly higher productivity growth relative to other kinds of FDI.

These and other researchers highlight the importance of supportive policies to create a suitable enabling environment as a means of attracting FDI and then enlarging the resulting spin-off of benefits to domestic producers. Areas in which policy can improve the prospects for attracting and benefiting from FDI include human capital (Borensztein, De Gregorio, and Lee 1998), developed financial markets (Alfaro and others 2004), and openness to trade (Balasubramanyam, Salisu, and Sapsford 1996).

Implementing Reform

Implementing a reform agenda to address the main constraints in the sectors with the most potential will require strong and sustained commitment from the top levels of government. Investors need to be confident that they are not under threat of government expropriation.

It is also important to determine whether a country has a comparative advantage in a given subsector. A simple way to do this is through domestic resource cost (DRC) ratios, which measure a country's comparative advantage in an industry (Bruno 1972; Bhagwati and Srinivasan 1980; Pack 1974, 1987). Because of its close links to the concept of effective protection, DRC ratios are also widely used as an index of economic efficiency in restrictionist trade regimes.

A DRC value less than 1 indicates that the cost of the domestic resources to produce a unit of the product is less than the potential foreign exchange earnings from exporting it. That is, the country has a comparative advantage, and there is a rationale for government to foster its exports. Similarly, a DRC value greater than 1 indicates that the cost of the domestic resources spent in producing a good for the domestic market is more than the foreign exchange spent importing it. That is, the country does not have a comparative advantage. If the government is supporting import-substituting policies for this product, it should discontinue them.

Table 8.4 Domestic Resource Cost Ratios in Ethiopia

Product	DRC ratio
Polo shorts	1.12
Leather loafers	0.78
Wooden chairs	4.16/1.73
Crown cork	−5.76
Wheat milling	0.93
Boxer briefs	Slightly below 1.0 (ERR)
Leather gloves	Above 1.0 (ERR)

Source: Global Development Solutions 2011.

For industries that pass the DRC test, whether for export or for import substitution, it would be useful to follow through with value chain studies to map constraints into policy recommendations and to identify exactly what it will take for the government to promote expansion of that industry. Table 8.4 shows the DRC ratio for each of the products for which the comparative value chain analysis was done. On the basis of the DRC ratio and the other criteria, the sector priorities, in decreasing order, are leather goods, agribusiness, apparel, metal, and wood products.

To attract both domestic and investors to light manufacturing in Ethiopia, the Ethiopian government should put in place a dedicated high-level implementation task team—together with transparent, inclusive, and professional processes—to develop and implement the reform program.

Botswana, Cape Verde, Malaysia, Mauritius, and Taiwan, China, did this at the outset of their economic transformation (Criscuolo and Palmade 2008). The combination of skills, access, and resources gave these teams the influence to steer an ambitious (yet focused) policy agenda in the face of opposing interests. Such teams have typically been charged with designing and updating the strategy, engaging and negotiating with potential leading investors, mobilizing the support of development partners, supporting and monitoring key government initiatives, keeping government leaders informed and committed, and breeding the next generation of leaders.

The policy approach in this chapter proposes active government support to the light manufacturing sector through means such as targeted policy reforms (especially in key input industries) as well as the provision of public goods such as "plug-and-play" industrial zones, technical assistance for entrepreneurs and workers, control of cattle diseases, and information on costs for potential foreign investors. This approach aims to foster competition, expand the capacity of all firms to compete, and level the playing field between types of firms. Because any active government policy creates the opportunity for rents or policy-induced profits, policy implementation should ensure that the beneficiaries are determined by market forces and not by the special interests of government officials or rent-seeking entrepreneurs.

The policy measures proposed here minimize the rent-seeking opportunities in implementation for five reasons:

- First, because the proposed approach and the sector-specific support are focused on sectors consistent with Ethiopia's latent comparative advantage, the extent of government support can be limited and rapidly scaled back as new information arrives.

- Second, to limit the extent and cost of rent seeking, policies should focus on providing public goods that provide widespread benefits.

- Third, the reform should begin with pilot studies and be continually revised and updated. In addition, implementation should be decentralized, to enhance proximity to the private sector, increase accountability, and foster competition among local governments.

- Fourth, the government must be ready to withdraw support for industries that fail to deliver anticipated gains.

- Fifth, one of the best ways the government can facilitate robust private sector growth is by maintaining a stable and conducive macroeconomic environment and by ensuring that natural resources are well managed.

Political Economy Issues

Economic policies have distributional impact. Most policy choices create distributional conflict—benefiting some and hurting others, vested interests in particular. The losers, if they are organized better and have stronger influence on the governments than the winners, can stop the reforms or force reversal of policies. Therefore, in designing and implementing reform programs, it is essential to identify the winners and losers and develop a strategy to reconcile the differences and elicit a favorable political response.

In the program recommended here, most of the policy actions would result in short- and longer-term gains for a large group of Ethiopians. In particular, improving trade logistics, developing industrial parks, enabling the use of land and cattle as collateral, controlling cattle disease, reducing the fuel tax and import duties on trucks and spare parts, and facilitating development of competitive domestic input industries would create large gains for a wide range of interest groups. Losses for some segments of business are expected from liberalization of imports of apparel inputs and leather (local producers of these products) and elimination of export bans on leather and cotton (local processors of these inputs) and price controls on some food products. But these measures would create a large number of winners too. For example, elimination of price caps on some food products would hurt some consumers but benefit producers, thus encouraging production and employment.

Even if both the number of beneficiaries and the scale of benefits would far exceed the number of losers and the magnitude of losses, as with the measures proposed in this report, favorable cost-benefit ratios do not guarantee smooth passage and implementation of the reform program. Mitigating this risk calls for consideration of several dimensions of program design:

- *Analyze the impact of proposed reforms.* The government should make every effort to mobilize support through public discussion and facilitate the creation of the broadest possible coalition in favor of reform. A gradual process of coalition building, especially among national and regional policy elites, was a key ingredient in China's eventual decision to adopt a "socialist market economy with Chinese characteristics," which represented a massive shift from the previous plan-oriented regime (Shirk 1993; Naughton 2008).

- *Structure reforms as a package designed to mobilize sufficiently broad support and counter potential opposition.* Phased implementation of some measures (import liberalization) might give potential losers time to adjust, thus reducing their losses and possibly muting their opposition to a reform package.

- *Leverage donor support.* Ethiopian manufacturing industries receive support from several donors. The German Development Bank (KfW), through the Engineering Capacity Building project, provides support to manufacturing industries similar to that of the World Bank's factory floor–level interventions to improve product quality and productivity by funding foreign technicians. It also supports market linkages, particularly to the EU and Germany. The Japan International Cooperation Agency provides support to about 30 medium and small enterprises and introduces the Kaizen system of continuous monitoring of productivity and quality improvement efforts, while the United Nations Industrial Development Organization provides support for benchmarking studies, particularly for the leather sector. The USAID supports market linkages for industries attempting to export to the U.S. market through fair trade participation and arranges direct buyer-seller linkages.

The proposed program of reforms is sharply focused and should therefore be fiscally manageable because its recommendations are few in number, are specific, and can be packaged and prioritized along the most promising subsectors (with the help of development partners). Table 8.5 lays out the costs and feasibility of this report's major recommendations. Although the proposed policies are designed to limit rent seeking, parts of the government may find ways to favor connected firms and extract rents from others. To avoid this, it is essential to secure and sustain the commitment of the top level of government to the growth and jobs agenda and to implement controls and incentives that will assure proper implementation.

Table 8.5 Estimated Fiscal Costs and Political Economy Feasibility of the Proposed Measures

Measure	Fiscal cost	Political economy feasibility
Fully liberalize the import and export of leather.	Import duties on leather represent less than $1 million a year.	The political risk is limited if the two measures are taken simultaneously: liberalized imports will help shoe manufacturers faced with leather shortages, while liberalized exports will ensure that tanneries get the world market price for their leather.
Give immediate and free access to foreign exchange (start with apparel and leather products).	Giving up the 1.5% foreign exchange fee charged to the importers of inputs for the apparel and leather industries would represent a fiscal loss of about US$500,000 a year. The apparel industry currently exports US$8 million, of which US$6 million represents reexport of imported materials (the fiscal loss on their imports would thus be US$120,000). Domestic apparel firms produce about US$50 million worth of apparel but import a much lower proportion (about 20 percent) because they rely on low-quality domestic textiles (the fiscal loss on their imports would thus be about US$150,000). The numbers in the leather product industry would be similar in magnitude.	Providing immediate access to currencies for these sectors should carry no cost, because these sectors have the potential to become very large net currency earners. This could be done by setting up a special reserve for the two sectors with controls to prevent use of the reserves for other purposes.
Establish green channels at customs (start with apparel and leather products).	The primary cost would be for technical assistance, which could be financed with the support of development partners.	This has been done many times before. There is no obvious political economy challenge.
Reduce the cost of letters of credit (start with apparel and leather products).	To be determined following in-depth diagnosis of costs; cost would be limited if the issue is related to some regulations or to the lack of competition in the banking sector.	Trade finance is a major source of income for commercial banks in Ethiopia. The banking industry would need to be persuaded and shown that the expected increase in the number and value of letters of credit would more than compensate for the reduction in fees; this could be first demonstrated in the context of the apparel and leather product subsectors, where the measure would have the most impact (the number of letters of credit issued by the banks in the apparel and leather industries is currently small).

Eliminate import tariffs on inputs for all firms (start with apparel, leather, and agribusiness).	The input components of individual subsectors are identified from data in the customs classification. Total import duty of the five subsectors amounted to Br 855 million (US$50 million) in 2009/10 (less than 2 percent of total tax revenue), of which more than 90 percent comes from the metal and garment industries. Removing import duties on inputs for the light manufacturing sector would not cost more than 2 percent of total tax revenue, which could easily be made up by imposing an excise tax if needed. Limiting the measure (at least initially) to the garment, leather, and agribusiness industries (where it would have the most benefits) would cost less than 1 percent of total tax revenues.	Import tariffs protect domestic suppliers. There is potential for domestic producers to lose out in the short run. The case needs to be made to show that all will benefit in the long run and that, if necessary, parallel interventions to support productivity improvements in domestic production will balance the loss to local farmers.
Reduce import tariffs on steel and investigate the feasibility of exploiting iron ore deposits (metal product value chain).	The reduction of import tariffs on steel will have a small fiscal impact.	To be studied.
Enable the use of land, machines, cattle, and outputs as collateral (all subsectors).	The cost would mostly be for technical assistance to improve the regulatory frameworks and collateral registries, which could be financed with the support of development partners.	This effort would leverage the successful reform that gave farmers the right to rent and inherit their land; the government could authorize the use of transferrable long-term lease rights, and such reforms could be first piloted in key locations.
Control cattle diseases.	The cost of combating ectoparasites is less than US$10 million (USAID). Promoting cross-breed cows and enabling the use of cattle as collateral would increase the incentives and financial means for farmers to access veterinary services.	The approach needs to be regional, as cattle cross borders.
Support entrepreneurial and worker skills.	The cost of providing management and technical training is high initially because expatriate trainers are needed. It declines significantly once local trainers become available. The cost of connecting clusters to utilities and providing them with the space to grow should, as for industrial parks, be a financing cost because small and medium enterprises should be charged only for recurring costs. Again, development partners might help with the financing. World-class, experienced providers of training should be leveraged, together with state-of-the-art methods of evaluation (randomized).	No obvious political economy challenge.

(table continued next page)

Table 8.5 *(continued)*

Measure	Fiscal cost	Political economy feasibility
Facilitate private sector investments in key agricultural input markets such as hybrid seeds and fertilizers.	Liberalizing the markets for key seeds and fertilizers could reduce the amounts spent by the government on credit guarantees, subsidies, and distribution.	This would go against the interests of the state-controlled companies that control the seeds and fertilizer markets; focusing or piloting such reforms on a few key crops where the reforms would have the most benefit would help to alleviate resistance.
Facilitate access to rural land to promote competitive and sustainable farms, cattle ranches, and wood plantations.	Facilitating access to land should generate (local) government revenues because the private sector should be asked to pay a land lease. The financing cost of connecting the land to the road network and utilities could be absorbed by the private investor in exchange for a reduction in the land lease. Carbon finance should be made available, since wood plantations on degraded land would help increase the forest area.	The government has already earmarked 3 million hectares of rural land for private investors. This is an extremely politically, socially, and environmentally sensitive topic that will require the government to rely on a fully transparent, inclusive, and highly professional process (principles for responsible investments in agriculture are being developed by the international community).
Extend the rural food safety nets to vulnerable urban populations as an alternative to price caps and export bans.	Removing the price caps on key food items could have fiscal implications because it may require targeted support to the poorest—for example, by extending the rural safety net to vulnerable urban populations (to be studied in detail).	Such reforms can be politically and socially sensitive. Lifting the export ban on cotton would benefit millions of farmers, but go against the short-term interest of textile mills. Extending the rural safety net to vulnerable urban populations would also need to be considered in the context of the food security agenda.
Develop plug-and-play industrial parks (would benefit all subsectors); start with one in Addis Ababa and conduct a feasibility study for one much closer to Djibouti (for example, in Dire Dawa).	Because tenants would cover the operating and maintenance costs of the parks, the fiscal cost is the financing cost of developing the park (about US$30 million for a 100-hectare park, which can accommodate 50,000 workers). As in China and Vietnam, private developers (including foreign ones) and banks could help to finance the parks.	The government already has a policy to facilitate access to industrial land for exporters in preferred industries (which include leather products, apparel, and agribusiness). The government is also engaged in developing industrial parks.
Facilitate the entry of leading investors along the value chains (start with apparel, leather, and agribusiness), once a critical mass of the measures listed above begins to take shape.		Strategy should be to design interventions that deliver benefits to the widest possible array of potential entrants in the subsector.

Source: Authors.
Note: Measures with significant short-term impact are listed first.

Notes

1. Investment data from http://www.unctad.org/sections/dite_dir/docs/wir11_fs_cn_en.pdf.
2. For Chinese data, see http://www.21tradenet.com/news_2011-1-11/83167.htm. For global trade total, see http://www.salisonline.org/market-research/trends-in-world-textile-and-clothing-trade-200910-edition/.

References

Alfaro, Laura, Areendam Chanda, Sebnem Kalemli-Ozcan, and Selin Sayek. 2004. "FDI and Economic Growth: The Role of Local Financial Markets." *Journal of International Economics* 64 (1): 89–112.

Alfaro, Laura, and Andrew Charlton. 2007. "Growth and the Quality of Foreign Direct Investment: Is All FDI Equal?" CEP Discussion Paper 830, Centre for Economic Performance, London School of Economics and Political Science, London.

Balasubramanyam, V. N., M. Salisu, and David Sapsford. 1996. "Foreign Direct Investment and Growth in EP and IS Countries." *Economic Journal* 106 (434): 92–105.

Bhagwati, Jagdish N., and T. N. Srinivasan. 1980. "Domestic Resource Costs, Effective Rates of Protection, and Project Analysis in Tariff-Distorted Economies." *Quarterly Journal of Economics* 94 (1): 205–09.

Borensztein, Eduardo, Jose De Gregorio, and J-W. Lee. 1998. "How Does Foreign Direct Investment Affect Economic Growth?" *Journal of International Economics* 45 (1): 115–35.

Bruno, Michael. 1972. "Domestic Resource Costs and Effective Protection: Clarification and Synthesis." *Journal of Political Economy* 80 (1): 16–33.

Chandra, Vandana, and Shashi Kolavalli. 2006. "Technology, Adaptation, and Exports: How Some Countries Got It Right." In *Technology, Adaptation and Exports,* ed., Vandana Chandra. Washington, DC: World Bank.

Criscuolo, Alberto, and Vincent Palmade. 2008. "Reform Teams." Viewpoint Note 318, World Bank, Washington, DC.

Du, Luosha, Ann Harrison, and Gary Jefferson. 2011. "Do Institutions Matter for FDI Spillovers? The Implications of China's 'Special Characteristics.'" NBER Working Paper 16767, National Bureau of Economic Research, Cambridge, MA.

Fafchamps, Marcel, and Simon Quinn. 2011. "Results from the Quantitative Firm Survey." Background paper (Light Manufacturing in Africa Study). Available online in Volume III at http://econ.worldbank.org/africamanufacturing. World Bank, Washington, DC.

Global Development Solutions. 2011. "The Value Chain and Feasibility Analysis; Domestic Resource Cost Analysis." Background paper (Light Manufacturing in Africa Study). Available online as Volume II at http://econ.worldbank.org/africamanufacturing. World Bank, Washington, DC.

Harrison, Ann, and Andres Rodríguez-Clare. 2010. "Trade, Foreign Investment, and Industrial Policy for Developing Countries." In *Handbook of Development Economics*, Vol. 5 of Dani Rodrik and Mark Rosenzweig eds., 4039–214. Amsterdam: North-Holland.

Hu, Albert, and Gary Jefferson. 2002. "FDI Impact and Spillover: Evidence from China's Electronic and Textile Industries." *World Economy* 25 (8): 1063–76.

Ministry of Finance and Economic Development. 2010. *Growth and Transformation Plan 2010/11–2014/15 Volume 1: Main Text.* Addis Ababa: Government of Ethiopia. http://www.mofed.gov.et/English/Resources/Documents/GTP%20English2.pdf.

National Bureau of Statistics of China. 2011. *China Statistical Yearbook 2010.* Beijing: China Statistics Press.

Naughton, Barry. 2008. "A Political Economy of China's Economic Transition." In *China's Great Economic Transformation,* ed. Loren Brandt and Thomas G. Rawski. Cambridge, U.K.: Cambridge University Press.

Pack, Howard. 1974. "The Employment-Output Trade-Off in LDC's: A Microeconomic Approach." *Oxford Economic Papers, New Series* 26 (3, November): 388–404.

———. 1987. *Productivity, Technology, and Industrial Development: A Case Study in Textiles.* Washington, DC: World Bank.

———. 1988. "Industrialization and Trade." In *Handbook of Development Economics,* Vol. 1, ed. Hollis Chenery and T. N. Srinivasan, ch. 9, 334–72. Amsterdam: North-Holland.

Shirk, Susan L. 1993. *The Political Logic of Economic Reform in China.* Berkeley: University of California Press.

USAID (U.S. Agency for International Development). 2008. "Success Story: Ethiopians Learning to Fight Ectoparasites." Financial Transactions and Reports Analysis, USAID, Washington, DC.

The Study's Objectives and Methods

This study aims to develop practical insights to help some African countries to become competitive in light manufacturing. This requires understanding why light manufacturing has been slow to grow in African countries but has taken off in other countries with similar development levels and investment climate indicators. It also requires understanding what strategies have been pursued by successful countries that not too long ago were at a similar development level and how such strategies could be adapted to African countries while fully recognizing their special circumstances.

The study aims to answer the following questions:

- Is there potential for light manufacturing in Africa? And if so, why, under generally similar constraints, are light manufacturing products that require fairly simple technologies and less-skilled labor produced and even exported by countries in all regions of the world except Africa, where they are not even produced for the domestic market?

- What are the critical constraints facing African firms? How have firms devised institutions and arrangements to circumvent some of these constraints?

- What are some practical policies that other countries have used to help firms to overcome the identified constraints, jump-start light manufacturing, and facilitate their transformation into modern economies?

Industry and Country Focus

The study focuses on five light manufacturing industries that are basic, simple, and labor intensive: agribusiness, leather goods, wood processing and wood products, simple metal products, and apparel.

Three countries in Africa were chosen for the study: Ethiopia, Tanzania, and Zambia. All three have been spared recent conflicts and experienced a degree of political stability over the years; such peace and stability are needed for the development of light manufacturing. Ethiopia was selected as the first case study for several reasons. It is among the largest African countries with relative political stability. Like many other African countries, it is landlocked. And its government has emphasized industrial growth as a pillar for the country's overall development. Tanzania has a long tradition in manufacturing, while Zambia has experienced similar problems with industrialization as many mineral-based African economies.

While many of the report's policy recommendations are based on Ethiopia, most of the policy issues addressed are common to many African countries. For this reason, the report does not always make a distinction between Ethiopia and Africa, although it does so where needed.

Two comparators that in the late 1980s had levels of development similar to those in Africa were selected from East Asia: China and Vietnam. Because China is the world's most competitive country in light manufacturing and a fierce competitor in Africa's domestic markets, it was chosen as the benchmark country for an in-depth study of the cost structure of production. China is relevant as a comparator because, when it emerged onto world markets, it had to compete in manufactured goods initially dominated by Hong Kong SAR, China; the Republic of Korea; Singapore; and Taiwan, China. When China started entering those labor-intensive sectors in 1979, it shared many of the investment climate constraints prevailing now in Ethiopia, which Ethiopia's low labor costs can offset if labor is the main input for production.

The sheer size of the Chinese economy makes it difficult for African countries to emulate its success. Vietnam is closer to African countries in size and level of development. Indeed, Vietnam does not have many of those favorable conditions that China has today. But Vietnamese products in many light manufacturing sectors compete with Chinese products in the domestic market, and some even compete in international markets because of Vietnam's low labor costs.

How producers in Vietnam cope with competition from China can provide insights for African countries by bringing into focus the competitiveness gap that African countries need to fill and the lessons they can learn to move forward. Vietnam turned out to be a good choice of comparator because many of its development characteristics lie between those of China and the three African countries (Fafchamps and Quinn 2011).

The decision to compare the three African countries with China and Vietnam was not based on the expectation that they can catch up right away. The conditions under which China and Vietnam initially started on their development path were different and have changed substantially since then. For example, both China and Vietnam started their reforms with a highly educated workforce and a stable political regime. And each country has its own resource endowments and comparative advantages.

But light manufacturing products coming out of Africa would have to compete with Asian products in today's market, and it is important to know what it would take for them to earn and keep market share. Moreover, a comparison of Africa with these Asian countries is certainly much more realistic than the traditional (implicit or explicit) comparison with production in the West, which has a longer and more complex history of development.

Methodology

The study draws on five analytical tools:

- New research based on the Enterprise Surveys
- Qualitative interviews with about 300 enterprises (both formal and informal of all sizes) by the study team in Ethiopia, Tanzania, and Zambia and in China and Vietnam, based on a questionnaire designed by Professor John Sutton of the London School of Economics[1]
- Quantitative interviews with about 1,400 enterprises (both formal and informal of all sizes) by the Centre for the Study of African Economies at Oxford University in the same countries, based on the questionnaire designed by Professor Marcel Fafchamps and Dr. Simon Quinn of Oxford University
- Comparative value chain and feasibility analysis based on in-depth interviews of about 300 formal medium enterprises in the same five countries, conducted by the consulting firm Global Development Solutions, Inc.
- A Kaizen study on the impact of managerial training for owners of small and medium enterprises. This training, delivered to about 550 entrepreneurs in Ethiopia, Tanzania, and Vietnam, was led by Japanese researchers from the Foundation for Advanced Studies on International Development and the National Graduate Institute for Policy Studies.

While these five data sources use different methodologies, their results are consistently similar, presenting a robust picture of the strengths and weaknesses of light manufacturing in Africa relative to East Asia.

Enterprise Survey Studies

Four studies were conducted based on World Bank Enterprise Surveys, which provide representative samples of each country's private sector. The surveys collect information on standard accounting measures of firm performance and many areas of each country's investment climate. Summaries of the broad survey results are posted on the Enterprise Surveys website (www.enterprise surveys.org).

Since 2006 the surveys have used almost identical questionnaires and identical sampling methodologies. They were conducted in two to five cities in each country and covered firms with more than five employees in manufacturing, construction, retail and wholesale services, hotels and restaurants, transport, storage, communications, and computer and related activities. The survey methodology is described in more detail and the survey data are available for download on the Enterprise Surveys website.

In the first study we used a large sample, containing more than 39,000 firms across 98 developing countries, to identify the most binding constraints on firm operations (Dinh, Mavridis, and Nguyen 2010). While each country faces a different set of constraints, the constraints also vary by firm characteristics, especially firm size.

In the second study we used firm-level data from 89 countries to examine the performance of African firms and the reasons for their disadvantages (Harrison, Lin, and Xu 2011). Our findings fail to confirm the perception that African manufacturing firms are not globally competitive and cannot be competitive in manufactured products. Formal manufacturing firms in Africa do not perform much worse than those in other countries at similar income levels, but they do exhibit structural problems: similar sales growth and higher labor growth, but slightly lower productivity and much lower export intensity and investment rates.

Given that the crucial issue of whether labor costs for formal firms are higher in Africa than in other countries at similar levels of development, we commissioned a third study to explore this question in greater depth, using data from the Enterprise Surveys, focusing on manufacturing (Clarke 2011a).

To make sure our study incorporates the latest findings in the literature concerning how the investment climate affects firm performance in Africa, we undertook a fourth study to review the empirical evidence on this issue (Clarke 2011b). A preponderance of evidence points to the importance of firm size.

Qualitative Survey

The study team interviewed firms with more than 300 owner-managers of enterprises (both formal and informal) of all sizes in all five countries, using the questionnaire designed by John Sutton as a guide. The objective was to learn about each firm's business, its owner-manager, and factors affecting business. Like the quantitative survey, this component of the study sought to go beyond the traditional investment climate surveys to probe the origin and capabilities of the firm's owner-manager, and other issues pertinent to the firm's operations and growth.

For the origins and development of capabilities to manage the firm, questions were asked about when and how the firm was set up, the original sources of capital, and where the idea for the business, as well as the know-how or technological knowledge, came from. Questions were also asked about how and why the firm's owner moved from one product to another and the role of family, relatives, or friends in the decision to make a product. For finance, we asked about the initial and later sources of financing to expand the business, reasons for not using bank financing, and the cost of financing. We also asked about staffing, including how hiring decisions were made. On inputs, we asked about sources of raw materials or inputs to produce the final product and the tariff or duty imposed on imported inputs. On market demand, we asked about the customers, the competition, and the government policies that affect the firm's competitiveness. We also asked about domestic competition and imports and about other types of constraints such as utilities and infrastructure.

Quantitative Survey

In many ways the quantitative survey mirrors the qualitative survey. It sought to quantify as far as possible the foregoing questions in the context of a firm survey and highlighted the contribution of factors that are different from those identified by the traditional investment climate assessments or Enterprise Surveys. This component of the study was carried out by the Centre for the Study of African Economies at Oxford University (Fafchamps and Quinn 2011).

The traditional Enterprise Surveys focus on the investment climate or external environment in which firms do business and identify policy-related constraints that need to be alleviated to promote a more conducive business environment. Many of these constraints relate to regulatory and bureaucratic hurdles, corruption, credit and interest rates, and availability of public services such as water, electricity, and roads that affect all types of firms. The presumption is that the capability of firms to grow is uniform and that they will respond positively to an improved investment climate. The policy implications of the traditional Enterprise Surveys are well known: the government can contribute to a more conducive business environment by alleviating these constraints. But the Enterprise Surveys cannot explain why producers in a country with the same investment climate produce some products and not others.

The quantitative survey focused on intrinsic differences among entrepreneurs, which can explain why some innovate, produce certain products, and prosper more than others in the same investment climate. The quantitative questionnaire asked the qualitative questions discussed above in a more rigorous framework suitable for advanced econometric techniques.

Comparative Value Chain and Feasibility Analysis

This analysis—by the Global Development Solutions of Reston, Virginia—complements the other elements of the study by providing a complete microeconomic framework to assess, for each product, the relative performance and potential of countries in productivity and costs and to identify some of the main factors contributing to low productivity and high costs. The comparative value chain and feasibility analysis highlights industry-specific factors (such as product market regulations and market failures) that tend to be overlooked by traditional cross-cutting approaches, such as the Enterprise Surveys (Global Development Solutions 2011).

The comparative value chain and feasibility analysis had the following objectives:

- Benchmark the competitiveness (productivity and costs) of a selected group of African countries against Asian competitors (China and Vietnam) in a representative sample of simple light manufacturing value chains
- Review the detailed breakdown of costs and productivity for each product and identify the main reasons for the productivity and cost gaps (such as policy and infrastructure issues, market failures, and a lack of social and human capital) done pro forma for the products not being produced
- Identify the most important and common constraints for each product and across the sample of products
- Generate insights into the possible practical solutions for addressing a critical mass of the identified constraints.

The detailed comparative value chain analysis benchmarked productivity and costs between firms in Ethiopia, Tanzania, and Zambia and the comparator countries, China and Vietnam. The analysis was based on interviews with more than 300 companies in these five countries and was conducted for specific products (representative of broader product categories or industries) to ensure the comparability of performance benchmarks, such as costs and productivity, between countries and to establish causality between external factors and firm performance through microeconomic analysis and company interviews. The conclusions can be generalized to broader product categories.

Six important and representative products were chosen for the value chain analyses: polo shirts, leather shoes (sheepskin loafers), wood chairs, crown corks (metal bottle caps), wheat flour, and processed milk. In addition, feasibility studies were conducted for men's boxer briefs, leather golf gloves, and padlocks.

Kaizen Training

In most African countries the private sector has a dual structure, with a large number of small indigenous enterprises and a small number of fairly large enterprises, often owned by foreigners or ethnic minorities or formerly owned by the government. These larger enterprises are, however, small and stagnant by the standards in other developing countries.

Drawing on the work at the National Graduate Institute for Policy Studies in Tokyo, Japan, this component of the study offered Kaizen training for micro and small entrepreneurs and assessed its short- and longer-term impact on firm performance. Kaizen is a self-help approach to efficiency improvements in organizations, also called lean manufacturing, which includes performance-based human resource management, continual analysis and refinement of quality control procedures, inventory management, and planning. Developed in the manufacturing sector in Japan, the Kaizen approach has evolved into what is now a standard set of modern management practices in Europe and the United States.

Most entrepreneurs of micro and small enterprises in developing countries are unaware of this approach to modern management. A major hypothesis is that imparting the basics of the Kaizen approach to entrepreneurs in micro and small enterprises will help their businesses to grow into small and medium enterprises. The National Graduate Institute for Policy Studies designed and implemented Kaizen training programs in Ethiopia, Tanzania, and Vietnam, so that participants could serve as a randomized pool of interviewees whose responses could be used to assess the impact of Kaizen training modules.

The impacts of Kaizen training were measured through an experiment with four major components: baseline surveys of enterprises, the planning and implementation of managerial training programs, the post-training surveys of enterprises, and impact evaluation of the managerial training programs based on the experimental data collected through the baseline and post-training surveys. ustry in Dar es Salaam, the engineering industry in Addis Ababa, the knitwear industry in Ha Tay Province, and the rolled steel product industry in Bac Ninh Province. The sample consisted of about 120 enterprises in the garment and textile industry in Dar es Salaam, about 100 enterprises in the engineering industry in Addis Ababa, about 120 enterprises in the knitwear industry in Ha Tay, and about 200 enterprises in the rolled steel products industry in Bac Ninh. Because the final results of the Kaizen impact evaluation will not be available until 2012, this study reports the interim results available in early 2011 (Sonobe, Sazuki, and Otsuka 2011).

Note

1. This questionnaire, and all the others mentioned in this paragraph, can be found in Fafchamps and Quinn 2011.

References

Clarke, George. 2011a. "Wages and Productivity in Manufacturing in Africa: Some Stylized Facts." Background paper (Light Manufacturing in Africa Study). Available online in Volume III at http://econ.worldbank.org/africamanufacturing. World Bank, Washington, DC.

———. 2011b. "Assessing How the Investment Climate Affects Firm Performance in Africa: Evidence from the World Bank's Enterprise Surveys." Background paper (Light Manufacturing in Africa Study). Available online in Volume III at http://econ.world bank.org/africamanufacturing. World Bank, Washington, DC.

Dinh, Hinh T., Dimitris Mavridis, and Hoa B. Nguyen. 2010. "The Binding Constraint on Firms' Growth in Developing Countries." Background paper (Light Manufacturing in Africa Study). Available online in Volume III at http://econ.worldbank.org /africamanufacturing. World Bank, Washington, DC.

Fafchamps, Marcel, and Simon Quinn. 2011. "Results from the Quantitative Firm Survey." Background paper (Light Manufacturing in Africa Study). Available online in Volume III at http://econ.worldbank.org/africamanufacturing. World Bank, Washington, DC.

Global Development Solutions. 2011. "The Value Chain and Feasibility Analysis; Domestic Resource Cost Analysis." Background paper (Light Manufacturing in Africa Study). Available online as Volume II at http://econ.worldbank.org/africamanufacturing. World Bank, Washington, DC.

Harrison, Ann E., Justin Y. Lin, and L. C. Xu. 2011. "Explaining Africa's (Dis) Advantage." Background paper (Light Manufacturing in Africa Study). Available online in Volume III at http://econ.worldbank.org/africamanufacturing. World Bank, Washington, DC.

Sonobe, Tetsushi, Aya Suzuki and Keijiro Otsuka. 2011. "Kaizen for Managerial Skills Improvement in Small and Medium Enterprises: An Impact Evaluation Study." Background paper (Light Manufacturing in Africa Study). Available online as Volume IV at http://econ.worldbank.org/africamanufacturing. World Bank, Washington, DC.

Index